ALCOHOL
AND TOBACCO
AMERICA'S DRUGS OF CHOICE

ALCOHOL AND TOBACCO
AMERICA'S DRUGS OF CHOICE

Sandra Alters

INFORMATION PLUS® REFERENCE SERIES
Formerly published by Information Plus, Wylie, Texas

GALE GROUP
THOMSON LEARNING

Detroit • New York • San Diego • San Francisco
Boston • New Haven, Conn. • Waterville, Maine
London • Munich

ALCOHOL & TOBACCO: AMERICA'S DRUGS OF CHOICE

Sandra Alters, *Author*

The Gale Group Staff:

Editorial: Ellice Engdahl, *Series Editor*; John F. McCoy, *Series Editor*; Charles B. Montney, *Series Editor*; Andrew Claps, *Series Associate Editor*; Jason M. Everett, *Series Associate Editor*; Michael T. Reade, *Series Associate Editor*; Heather Price, *Series Assistant Editor*; Teresa Elsey, *Editorial Assistant*; Debra M. Kirby, *Managing Editor*; Rita Runchock, *Managing Editor*

Image and Multimedia Content: Barbara J. Yarrow, *Manager, Imaging and Multimedia Content*; Robyn Young, *Project Manager, Imaging and Multimedia Content*

Indexing: Susan Kelsch, *Indexing Supervisor*

Permissions: Margaret Chamberlain, *Permissions Specialist*; Maria Franklin, *Permissions Manager*

Product Design: Michelle DiMercurio, *Senior Art Director and Product Design Manager*; Michael Logusz, *Cover Art Designer*

Production: Evi Seoud, *Assistant Manager, Composition Purchasing and Electronic Prepress*; NeKita McKee, *Buyer*; Dorothy Maki, *Manufacturing Manager*

Cover photo © Digital Stock.

ISBN 0-7876-5103-6 (set)
ISBN 0-7876-5391-8 (this volume)
ISSN 1536-5239 (this volume)
Printed in the United States of America
10 9 8 7 6 5 4 3 2 1

TABLE OF CONTENTS

Alcohol and tobacco are drugs. This chapter introduces concepts important to the discussion of these substances, offering a definition and classification of drugs and an introduction to abuse and addiction.

People have used alcohol, tobacco, and caffeine for hundreds of years. This chapter provides a history of each, revealing the degree to which use of these substances has become entwined with human societies.

Alcohol is a very commonly used drug. This chapter discusses the consumption of alcohol in America statistically and describes the short- and long-term effects of the use of alcohol. The use of alcohol during pregnancy, the interaction of alcohol with other drugs, the health dangers and benefits of using alcohol, and accidents and crimes related to the use of alcohol are also explored.

This chapter begins by defining alcoholism and alcohol abuse. The sociological, psychological, behavioral, and genetic causes of these disorders are considered. Also discussed are the costs and treatment of alcoholism and alcohol abuse.

Tobacco was used in the United States before the arrival of Columbus and is widely used today. This chapter discusses the properties of nicotine, statistics on tobacco use, the health consequences of tobacco use and secondhand smoke, the movement to ban smoking, and quitting smoking.

A number of surveys are used to track substance use among young people. This chapter outlines the results of these surveys, describing the attitudes of youth toward alcohol and tobacco, youth alcohol and tobacco use, and the availability of alcohol and tobacco to youth.

Alcohol and tobacco are a big business. Production, sales, and consumption of alcohol and tobacco are described statistically in this chapter. Special attention is given to the advertising of alcohol and tobacco products.

The federal government's taxes and regulations on alcohol and tobacco are outlined in this chapter. Court cases seeking damages from the tobacco companies, including the Master Settlement Agreement, are also discussed.

Caffeine is a drug, just like alcohol and tobacco. This chapter explores the sources of caffeine, average caffeine consumption, and the physical and health effects of caffeine, including the question of addiction. Additionally, the process of decaffeination is explained.

This chapter takes a look at people's use of and opinions on alcohol and tobacco. Issues raised include family problems caused by alcohol use, underage drinking, quitting smoking, limiting smoking in public places, responsibility for health damage caused by smoking, and government regulation of tobacco.

PREFACE

Alcohol and Tobacco: America's Drugs of Choice is one of the latest volumes in the Information Plus Reference Series. Previously published by the Information Plus company of Wylie, Texas, the Information Plus Reference Series (and its companion set, the Information Plus Compact Series) became a Gale Group product when Gale and Information Plus merged in early 2000. Those of you familiar with the series as published by Information Plus will notice a few changes from the 1999 edition. Gale has adopted a new layout and style that we hope you will find easy to use. Other improvements include greatly expanded indexes in each book, and more descriptive tables of contents.

While some changes have been made to the design, the purpose of the Information Plus Reference Series remains the same. Each volume of the series presents the latest facts on a topic of pressing concern in modern American life. These topics include today's most controversial and most studied social issues: abortion, capital punishment, care for the elderly, crime, health care, the environment, immigration, minorities, social welfare, women, youth, and many more. Although written especially for the high school and undergraduate student, this series is an excellent resource for anyone in need of factual information on current affairs.

By presenting the facts, it is Gale's intention to provide its readers with everything they need to reach an informed opinion on current issues. To that end, there is a particular emphasis in this series on the presentation of scientific studies, surveys, and statistics. These data are generally presented in the form of tables, charts, and other graphics placed within the text of each book. Every graphic is directly referred to and carefully explained in the text. The source of each graphic is presented within the graphic itself. The data used in these graphics is drawn from the most reputable and reliable sources, in particular the various branches of the U.S. government and major independent polling organizations. Every effort has been made to secure the most recent information available. The reader should bear in mind that many major studies take years to conduct, and that additional years often pass before the data from these studies is made available to the public. Therefore, in many cases the most recent information available in 2001 dated from 1998 or 1999. Older statistics are sometimes presented as well, if they are of particular interest and no more recent information exists.

Although statistics are a major focus of the Information Plus Reference Series, they are by no means its only content. Each book also presents the widely held positions and important ideas that shape how the book's subject is discussed in the United States. These positions are explained in detail and, where possible, in the words of their proponents. Some of the other material to be found in these books includes: historical background; descriptions of major events related to the subject; relevant laws and court cases; and examples of how these issues play out in American life. Some books also feature primary documents, or have pro and con debate sections giving the words and opinions of prominent Americans on both sides of a controversial topic. All material is presented in an even-handed and unbiased manner; the reader will never be encouraged to accept one view of an issue over another.

HOW TO USE THIS BOOK

Beyond those prescription and over-the-counter medicines found in pharmacies and drugstores, there are also a few legal drugs that are found commonly in everyday life. Alcohol, tobacco, and caffeine are all legal substances that can affect a person's mood and/or physiology. These three substances are all readily available, generally affordable, and economically important. Depending on time, place, and circumstance, they can be more or less socially acceptable. This book provides an overview of all three substances, including their history, health impact, addictive nature, and potential for abuse. Also discussed are the political and economic ramifications of alcohol, tobacco, and caffeine; their use by youth; and public opinions about them.

Alcohol and Tobacco: America's Drugs of Choice consists of ten chapters and three appendices. Each of the chapters is devoted to a particular aspect of alcohol, tobacco, and/or caffeine. For a summary of the information covered in each chapter, please see the synopses provided in the Table of Contents at the front of the book. Chapters generally begin with an overview of the basic facts and background information on the chapter's topic, then proceed to examine sub-topics of particular interest. For example, Chapter 6: Alcohol, Tobacco, and Youth begins with a discussion of several of the most important surveys of alcohol and tobacco use among American youth. It then examines the way American children and teenagers view alcohol and tobacco, the prevalence of underage use of these substances, and drinking and driving among teenagers. Also covered are the health consequences of early tobacco use, the availability of alcohol and tobacco to minors, and the use of these substances among college students and other young adults. Readers can find their way through a chapter by looking for the section and sub-section headings, which are clearly set off from the text. Or, they can refer to the book's extensive index if they already know what they are looking for.

Statistical Information

The tables and figures featured throughout *Alcohol and Tobacco: America's Drugs of Choice* will be of particular use to the reader in learning about this issue. These tables and figures represent an extensive collection of the most recent and important statistics on alcohol, tobacco, and caffeine, as well as related issues—for example, graphics in the book cover the amount of alcoholic beverages consumed per capita by American citizens over the past several decades, alcohol involvement in fatal automobile accidents, the estimated economic costs of alcohol abuse, underage use of tobacco, the benefits of smoking cessation, and comparisons of caffeine amounts in different foods and beverages. Gale believes that making this information available to the reader is the most important

way in which we fulfill the goal of this book: to help readers understand the issues and controversies surrounding alcohol, tobacco, and caffeine in the United States and reach their own conclusions.

Each table or figure has a unique identifier appearing above it, for ease of identification and reference. Titles for the tables and figures explain their purpose. At the end of each table or figure, the original source of the data is provided.

In order to help readers understand these often complicated statistics, all tables and figures are explained in the text. References in the text direct the reader to the relevant statistics. Furthermore, the contents of all tables and figures are fully indexed. Please see the opening section of the index at the back of this volume for a description of how to find tables and figures within it.

In addition to the main body text and images, *Alcohol and Tobacco: America's Drugs of Choice* has three appendices. The first is the Important Names and Addresses directory. Here the reader will find contact information for a number of government and private organizations that can provide information on alcohol, tobacco, and/or caffeine. The second appendix is the Resources section, which can also assist the reader in conducting his or her own research. In this section, the author and editors of *Alcohol and Tobacco: America's Drugs of Choice* describe some of the sources that were most useful during the compilation of this book. The final appendix is the index. It has been greatly expanded from previous editions, and should make it even easier to find specific topics in this book.

COMMENTS AND SUGGESTIONS

The editors of the Information Plus Reference Series welcome your feedback on *Alcohol and Tobacco: America's Drugs of Choice*. Please direct all correspondence to:

Editor
Information Plus Reference Series
27500 Drake Rd.
Farmington Hills, MI, 48331-3535

ACKNOWLEDGEMENTS

Illustrations appearing in Alcohol and Tobacco 2001 *were received from the following sources:*

Alcohol Alert: no. 30 PH 359, October 1995

Alcohol Health & Research World: "Alcohol Hangover" by Robert Swift and Dena Davidson, vol. 22, no. 1, 1998; "A Behavioral-Genetic Perspective on Children of Alcoholics" by Matt McGue, vol. 21, no. 3, 1997; "Distinguishing Differences Between Type I and Type II Alcoholism" by C. Robert Cloninger et al., v. 20, 1996. Reproduced by permission of the author; "Gender and Alcoholic Subtypes," by Frances K. Del Boca and Michie N. Hesselbrock, vol. 20, no. 1, 1996; "Late Life Drinking Behavior: The Influence of Personal Characteristics, Life Context, and Treatment" by Penny L. Brennan and Rudolf H. Moos, vol. 20, no. 3, 1996; "Risks and Benefits of Alcohol Use over the Life Span" by Mary C. Dufour, vol. 20, no. 3, 1996; vol. 17, no. 1, 1993 (updated October, 2000)

The American Journal of Psychiatry: "Prevalence and Pair Resemblance for Lifetime Alcohol Abuse and Dependence Among 1,514 Male Twin Pairs," by Carol A. Prescott and Kenneth S. Kendler, v. 156, 1999. Copyright (c) 1999 the American Psychiatric Association. Reprinted by permission of the publisher and the respective authors.

American Lung Association: Tobacco Control Comparison maps 3/01

Bureau of Justice Statistics: *Alcohol and Crime,* 1998; *Substance Abuse and Treatment, State and Federal Prisoners, 1997,* 1999

Bureau of Labor Statistics

Center for Science in the Public Interest: *Alcohol Policies Project,* 2000

Centers for Disease Control and Prevention: *National Health Interview Survey, 2000,* 2000; "Tobacco Use—United States,

1900–1999" in *Morbidity and Mortality Weekly Report,* vol. 48, no. 43, November 5, 1999; "Youth Risk Behavior Surveillance—United States, 1999" in *Morbidity and Mortality Weekly Report,* vol. 49, no. SS-5, June 9, 2000; "Youth Tobacco Surveillance, United States, 1998–1999" in *Morbidity and Mortality Weekly Report,* vol. 49, no. SS-10, October 13, 2000

Congressional Research Service, The Library of Congress: *Tobacco Master Settlement Agreement (1998): Overview and Issues for the 106th Congress,* 1999

Gallup News Service: *Nine of Ten Americans View Smoking as Harmful* by David W. Moore, October 7, 1999; *One in Six Americans Admit Drinking Too Much* by Wendy W. Simmons, December 4, 2000; *Smoking in Restaurants Frowned on by Many Americans* by Lydia Saad, November 29, 2000; *Tobacco Pact Generates Mixed Reactions* by Frank Newport, June 28, 1997

Indiana Prevention Resource Center at Indiana University: *Blood Alcohol Concentration Limits for Enforcement of Impaired Driving Laws —U.S. States —2001,* 2001

Institute for Social Research, University of Michigan, and the National Institute on Drug Abuse, U.S. Department of Health and Human Services: *Monitoring the Future: National Results on Adolescent Drug Use— Overview of Key Findings, 2000* by Lloyd D. Johnston, Patrick M. O'Malley, and Jerald G. Bachman, 2001; *Monitoring the Future: National Survey Results on Drug Use, 1975–1999: Volume 1: Secondary School Students 1999,* by Lloyd D. Johnston, Patrick M. O'Malley, and Jerald G. Bachman, 2000

Journal of the American College of Cardiology: "Light-to-Moderate Alcohol Consumption and Mortality in the Physicians' Health Study Enrollment Cohort" by J. Michael Gaziano et al., vol. 35, no. 1, 2000

National Center for Health Statistics: *Health, United States, 2000,* 2000

National Highway Traffic Safety Administration, National Center for Statistics and Analysis: *Traffic Safety Facts 1999—Alcohol,* 2000

National Institute of Drug Abuse

National Institute on Alcohol Abuse and Alcoholism: *Apparent Per Capita Alcohol Consumption: National, State, and Regional Trends, 1977–98* by Thomas M. Nephew, Gerald D. Williams, Frederick S. Stinson, Kim Nguyen, and Mary C. Dufour, 2000; "Effects of Alcohol on Areas of the Brain"; "The Path Alcohol Takes After Consumption"; *Trends in Alcohol-Related Morbidity Among Short-Stay Community Hospital Discharges, United States, 1979–97* by Christine C. Whitmore, Frederick S. Stinson, and Mary C. Dufour, 1999

The New England Journal of Medicine: "Alcohol Consumption and Mortality Among Middle-Aged and Elderly U.S. Adults," by Michael J. Thun, Richard Penn, Alan D. Lopez, Jane H. Monaco, Jane Henley, Clark W. Heath and Richard Doll, vol. 337, no. 24, December 11, 1997

Office of Technology Assessment, Congress of the United States: *Substance Abuse and Addiction,* 1994

PRIDE Surveys: *PRIDE Questionnaire Report: 1999–2000 National Summary, Grades 6 Through 12,* 2000

Substance Abuse and Mental Health Services Administration: *National Household Survey on Drug Abuse: Population Estimates 1998,* 1999; *Summary of Findings from the 1998 National Household Survey on Drug Abuse,* 1999; *Summary of Findings from the 1999 National Household Survey on Drug Abuse,* 2000

University of Minnesota: *Youth Access to Alcohol Survey: Summary Report* by E.M. Harwood, A.C. Wagenaar, and K.M. Zander, 1998. Reproduced by permission of E.M. Harwood.

U.S. Department of Agriculture: *Agricultural Statistics 2000,* National Agricultural Statistics Service, 2000; *Tobacco Situation and Outlook Report,* Economic Research Service, April 2001; *Tobacco Situation and Outlook Yearbook,* Economic Research Service, December 2000

U.S. Department of Commerce, Bureau of the Census: "State Government Tax Collections: 1999," 2000

U.S. Department of Health and Human Services: *10th Special Report to the U.S. Congress on Alcohol and Health,* June 2000; *The Health Benefits of Smoking Cessation: A Report of the Surgeon General,* 1990

U.S. Department of Health and Human Services, Centers for Disease Control and Prevention, and National Center for Chronic Disease

Prevention and Health Promotion, Office on Smoking and Health: *Investment in Tobacco Control: State Highlights–2000,* 2001

U.S. Department of the Treasury, Bureau of Alcohol, Tobacco, and Firearms: "Alcohol, Tobacco, and Firearms Tax Collections, Cumulative Summary Fourth Quarter Fiscal Year," 1999

CHAPTER 1

WHAT ARE "LEGAL DRUGS"?

DEFINITION AND REGULATORY AGENCIES

Drugs are nonfood chemicals that alter the way a person thinks, feels, functions, or behaves. Legal drugs are drugs whose sale, possession, and use as intended are not forbidden by law. However, the United States Drug Enforcement Administration (DEA) controls the use of legal psychoactive (mood- or mind-altering) drugs because of their potential for abuse. These drugs, which include narcotics, depressants, and stimulants, are available only with a prescription and are called "controlled substances."

The goal of the DEA is to ensure that controlled substances are readily available for medical use, while preventing their illegal sale and abuse. This agency works toward accomplishing its goal by requiring persons and businesses that manufacture, distribute, prescribe, and dispense controlled substances to register with the DEA. Registrants must abide by a series of requirements relating to drug security, records accountability, and adherence to standards.

The U.S. Food and Drug Administration (FDA) also plays a role in drug control. This agency regulates the manufacture and marketing of prescription and nonprescription drugs, requiring the active ingredients in a product to be safe and effective before allowing the drug to be sold.

Alcohol and tobacco are monitored and specially taxed by the Bureau of Alcohol, Tobacco, and Firearms (ATF). The alcohol program of this governmental agency regulates the qualification and operations of distilleries, wineries, and breweries. Additionally, it tests alcoholic beverages to ensure that their regulated ingredients are within legal limits and monitors labels for misleading information. The ATF tobacco program screens applicants who wish to manufacture, import, or export tobacco products.

Five Categories of Substances

Drugs may be classified into five categories:

- Depressants, including alcohol and tranquilizers. These substances slow down the activity of the nervous system. They produce sedative (calming) and hypnotic (trancelike) effects as well as drowsiness. If taken in large doses, depressants can cause intoxication (drunkenness).

- Hallucinogens, including marijuana, PCP (phencyclidine), and LSD (lysergic acid diethylamide). Hallucinogens produce abnormal and unreal sensations such as seeing distorted and vividly colored images. Hallucinogens can produce frightening psychological responses such as anxiety, depression, and the feeling of losing control of one's mind.

- Narcotics, including heroin and opium, from which morphine and codeine are derived. Narcotics are drugs that alter the perception of pain and induce sleep and euphoria (an intense feeling of well-being; a "high").

- Stimulants, including caffeine, nicotine, cocaine, amphetamines, and methamphetamines. These substances speed up the processing rate of the central nervous system. They can reduce fatigue, elevate mood, increase energy, and help people stay awake. In large doses, stimulants can cause irritability, anxiety, sleeplessness, and even psychotic behavior. Caffeine is the most commonly used stimulant in the world.

- Other compounds, including anabolic steroids and inhalants. Anabolic steroids are a group of synthetic substances that are chemically related to testosterone and are promoted for their muscle-building properties. Inhalants are solvents and aerosol products that produce vapors having psychoactive effects. These substances dull pain and can produce euphoria.

Table 1.1 provides an overview of alcohol, nicotine, and selected other psychoactive substances.

TABLE 1.1

Commonly Abused Drugs

Substance: Category and Name	Examples of *Commercial* and Street Names	DEA Schedule*/ How Administered**	Intoxication Effects/Potential Health Consequences
Depressants			
alcohol	beer, wine, hard liquor	not scheduled/swallowed	reduced pain and anxiety; feeling of well-being; lowered inhibitions; slowed pulse and breathing; lowered blood pressure; poor concentration/confusion, fatigue; impaired coordination, memory, judgment; respiratory depression and arrest, addiction
barbiturates	*Amytal, Nembutal, Seconal, Phenobarbital;* barbs, reds, red birds, phennies, tooies, yellows, yellow jackets	II, III, V/injected, swallowed	
benodiazepines (other than flunitrazepam)	*Ativan, Halcion, Librium, Valium, Xanax;* candy, downers, sleeping pills, tranks	IV/swallowed	Also, for barbiturates—sedation, drowsiness/depression, unusual excitement, fever, irritability, poor judgment, slurred speech, dizziness
flunitrazepam***	*Rohypnol;* forget-me pill, Mexican Valium, R2, Roche, roofies, roofinol, rope, rophies	IV/swallowed, snorted	for benzodiazepines—sedation, drowsiness/dizziness
			for flunitrazepam—visual and gastrointestinal disturbances, urinary retention, memory loss for the time under the drug's effects
GHB***	*gamma-hydroxybutyrate;* G, Georgia home boy, grievous bodily harm, liquid ecstasy	under consideration/swallowed	for GHB drowsiness, nausea/vomiting, headache, loss of consciousness, loss of reflexes, seizures, coma, death
methaqualone	*Quaalude, Sopor, Parest;* ludes, mandrex quad, quay	I/injected, swallowed	for methaqualone—euphorial/depression, poor reflexes, slurred speech, coma
Cannabinoids (Hallucinogens)			
hashish	boom, chronic, gangster hash, hash oil, hemp	I/swallowed, smoked	euphoria, slowed thinking and reaction time, confusion, impaired balance and coordination/cough, frequent respiratory infections; impaired memory and learning; increased heart rate, anxiety; panic attacks; tolerance, addiction
marijuana	blunt, dope, ganja, grass, herb, joints, Mary Jane, pot, reefer, sinsemilla, skunk, weed	I/swallowed, smoked	
Dissociative Anesthetics (Hallucinogens)			
ketamine	*Ketalar SV;* cat Valiums, K, Special K, vitamin K	III/injected, snorted, smoked	increased heart rate and blood pressure, impaired motor function/memory loss numbness; nausea/vomiting
			Also, for ketamine—at high doses, delirium, depression, respiratory depression and arrest
PCP and analogs	*phencyclidine;* angel dust, boat hog, love boat, peace pill	I, II/injected, swallowed, smoked	for PCP and analogs—possible decrease in blood pressure and heart rate, panic, aggression, violence/loss of appetite, depression
Hallucinogens			
LSD	*lysergic acid diethylamide;* acid, blotter, boomers, cubes, microdot, yellow sunshines	I/swallowed, absorbed through mouth tissues	altered states of perception and feeling; nausea/chronic mental disorders, persisting perception disorder (flashbacks)
mescaline	buttons, cactus mesc, peyote	I/swallowed, smoked	Also for LSD and mescaline—increased body temperature, heart rate, blood pressure; loss of appetite, sleeplessness, numbness, weakness, tremors
psilocybin	magic mushroom, purple passion, shrooms	I/swallowed	for psilocybin—nervousness, paranoia

DRUGS DISCUSSED IN THIS BOOK

This book focuses on three substances widely used throughout the world—alcohol, tobacco, and caffeine. Not only are alcohol, tobacco, and caffeine legal, relatively affordable, and more or less socially acceptable (depending on time, place, and circumstance), but they are also important economic commodities. Industries exist to produce, distribute, and sell these products, creating jobs and income, and contributing to economic well-being. The possible government regulation of alcohol and tobacco raises significant economic and political issues.

WHAT ARE ABUSE AND ADDICTION?

Scientists do not know why some people who use addictive substances become addicted and others do not. Results of numerous studies of identical and fraternal

TABLE 1.1
Commonly Abused Drugs [CONTINUED]

Substance: Category and Name	Examples of Commercial and Street Names	DEA Schedule*/ How Administered**	Intoxication Effects/Potential Health Consequences
Opioids and Morphine Derivatives (Narcotics)			
codeine	Empirin with Codeine, Fiorinal with Codeine, Robitussin A-C, Tylenol with Codeine; Captain Cody, Cody, schoodboy; (with glutethimide) doors & fours, loads, pancakes and syrup	II, III, IV/injected swallowed	pain relief, euphoria, drowsiness/respiratory depression and arrest, nausea, confusion constipation, sedation, unconsciousness, coma, tolerance, addiction Also, for codeine—less analgesia, sedation, and respiratory depression than morphine
fentanyl	Actiq, Duragesic, Sublimaze; Apache, China girl, China white, dance fever, friend, goodfella, jackpot, murder 8, TNT, Tango and Cash	II/injected, smoked, snorted	
heroin	diacetylmorphine; brown sugar, dope, H, horse, junk, skag, skunk, smack, white horse	I/injected, smoked, snorted	
morphine	Roxanol, Duramorph; M, Miss Emma, monkey, white stuff	II, III/injected, swallowed, smoked	
opium	laudanum, paregoric; big O, black stuff, block, gum, hop	II, III, V/swallowed, smoked	
amphetamine	Biphetamine, Dexedrine; bennies, black beauties, beauties, crosses, hearts, LA turnaround, speed, truck drivers, uppers	II/injected, swallowed, smoked, snorted	feelings of exhilaration, energy, increased mental alertness/rapid or irregular heart beat; reduced appetite, weight loss, heart failure Also, for amphetamine—rapid breathing hallucinations/tremor, loss of coordination; irritability, anxiousness, restlessness, delirium, panic, paranoia, impulsive behavior, aggressiveness, tolerance, addiction
cocaine	Cocaine hydrochloride; blow, bump, C, candy, Charlie coke, crack, flake, rock, snow, toot	II/injected smoked, snorted	for cocaine—increased temperature/chest pain, respiratory failure, nausea, abdominal pain, strokes, seizures, headaches, malnutrition
MDMA (methylenediosy-methamphetamine	DOB, DOM, MDA; Adam, clarity, ecstasy, Eve, lover's speed, peace, STP, X, XTC	I/swallowed	for MDMA—mild hallucinogenic effects, increased tactile sensitivity, empathic feelings, hyperthermial/impaired memory and learning
methamphetamine	Desoxyn; chalk, crank, crystal, fire, glass, go fast, ice, meth, speed	II/injected, swallowed, smoked,	for methamphetamine—aggression, violence, psychotic behavior/memory loss, cardiac and neurological damage; impaired memory and learning, tolerance, addiction
methylphenidate	Ritalin; JIF, MPH, R-ball, Skippy, the smart drug, vitamin R	II/injected, swallowed, snorted	for methylphenidate—increase or decrease in blood pressure, psychotic episodes/digestive problems, loss of appetite weight loss
nicotine	bidis, chew, cigars, cigarettes, smokeless tobacco, snuff,	not scheduled/smoked, snorted, taken in snuff and spit tobacco	for nicotine—tolerance, addiction; additional effects attributable to tobacco exposure-adverse pregnancy outcomes, chronic lung disease, stroke, cancer

(non-identical) twins and families with histories of substance abuse and addiction indicate that there is probably a genetic component to addiction. To date, however, researchers have not identified specific genes that would identify persons who are at risk of becoming addicted.

Research and treatment experts have identified three general levels of interaction with drugs: use, abuse, and dependence (or addiction). In general, abuse involves a compulsive use of a substance and impaired social or occupational functioning. Dependence (addiction) includes these traits, plus evidence of physical tolerance (a need to take increasingly higher doses to achieve the same effect) or withdrawal symptoms when use of the drug is stopped.

The progression from use to dependence is very complex, as are the abused substances themselves. Researchers have found no standard boundaries between using a substance, abusing a substance, and being addicted to a

TABLE 1.1

Commonly Abused Drugs [CONTINUED]

Substance: Category and Name	Examples of Commercial and Street Names	DEA Schedule*/ How Administered**	Intoxication Effects/Potential Health Consequences
Other Compounds			
anabolic steroids	Anadrol, Oxandrin, Durabolin, Depo-Testosterone, Equipoise; roids, juice	III/injected, swallowed, applied to skin	no intoxication effects/hypertension, bloc clotting and cholesterol changes, liver cysts and cancer, kidney cancer, hostility and aggression, acne; adolescents, premature stoppage of growth; in males, prostate cancer, reduced sperm production, shrunken testicles, breast enlargement; in females, menstrual irregularities, development of beard and other masculine characteristics
inhalants	Solvents (paint thinners, gasoline, glues), gases (butane, propane, aerosol propellants, nitrous oxide), nitrites (isoamyl, isobutyl, cyclohexyl); laughing gas, poppers, snappers, whippets	not scheduled/inhaled through nose or mouth	stimulation, loss of inhibition; headache; nausea or vomitint; slurred speech, loss of motor coordination; wheezing/ unconsciousness, cramps, weight loss, muscle weakness, depression, memory impairment, damage to cardiovascular and nervous systems, sudden death

* Schedule I and II drugs have a high potential for abuse. They require greater storage security and have a quota on manufacturing, among other restrictions. Schedule I drugs are available for research only and have no approved medical use; Schedule II drugs are available only by prescription (unrefillable) and require a form for ordering. Schedule III and IV drugs are available by prescription, may have five refills in 6 months, and may be ordered orally. Most Schedule V drugs are available over the counter.
** Taking drugs by injection can increase the risk of infection through needle contamination with staphylococci, HIV, hepatitis, and other organisms.

SOURCE: "Commonly abused drugs," National Institute of Drug Abuse, accessed on April 23, 2001 at http://www.nida.nih.gov/DrugsofAbuse.html

substance. They believe these lines vary widely from substance to substance and from individual to individual.

Physiological, Psychological, and Sociocultural Factors

Some researchers maintain that the principal causes of substance use are external social influences, such as peer pressure, while substance abuse and/or dependence result primarily from internal psychological and physiological needs and pressures, including inherited tendencies. Additionally, psychoactive drug use at an early age may be a risk factor (a characteristic that increases likelihood) for subsequent dependence.

Physically, mood-altering substances affect brain processes. Most drugs that are abused stimulate the reward or pleasure centers of the brain by causing the release of dopamine, a neurotransmitter, or chemical in the brain that relays messages from one nerve cell to another.

Psychologically, a person may become dependent on a substance because it relieves pain, offers escape from real or perceived problems, or makes the user feel more relaxed or confident in certain social settings. A successful first use of a substance may reduce the user's fear of the drug and thus lead to continued use and even dependence.

Socially, substance use may be widespread in some groups or environments. The desire to belong to a special group is a very strong human characteristic, and those who use one or more substances may become part of a subculture that encourages and promotes use. An individ-

ual may be influenced by one of these groups to start using a substance, or he or she may be drawn to such a group after starting use somewhere else. In addition, a person—especially a young person—may not have access to alternative rewarding or pleasurable groups or activities that do not include substance use.

Figure 1.1 illustrates some relationships between physiological, psychological, and sociocultural factors that influence drinking and drinking patterns. Constraints (inhibitory factors) and motivations influence drinking patterns. In turn, drinking patterns influence the relationship between routine activities related to drinking and acute (immediate) consequences of drinking.

Definitions of Abuse and Dependence

Two texts provide the most commonly used medical definitions of substance abuse and dependence. The *Diagnostic and Statistical Manual of Mental Disorders* (DSM) is published by the American Psychiatric Association. The *International Classification of Diseases* (ICD) is published by the World Health Organization (WHO). While the definitions of dependence in the two manuals are almost identical, the definitions of abuse are not.

THE DSM DEFINITION OF ABUSE. The text revision of the fourth edition of the DSM (DSM-IV-TR) defines abuse as an abnormal pattern of recurring use that leads to "significant impairment or distress," marked by one or more of the following in a 12-month period:

FIGURE 1.1

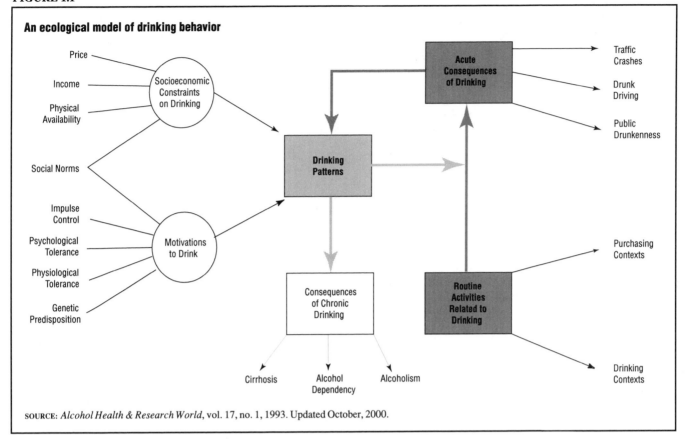

An ecological model of drinking behavior

SOURCE: *Alcohol Health & Research World*, vol. 17, no. 1, 1993. Updated October, 2000.

• Failure to fulfill major obligations at home, school, or work (for example, repeated absences, poor performance, or neglect).

• Use in hazardous or potentially hazardous situations, such as driving a car or operating a machine while impaired.

• Legal problems, such as arrest for disorderly conduct while under the influence of the substance.

• Continued use in spite of social or interpersonal problems caused by the use of the substance, such as fights or family arguments.

THE ICD DEFINITION OF "HARMFUL USE." The tenth edition of the ICD (ICD-10) uses the term "harmful use" rather than abuse. It defines harmful use as "a pattern of psychoactive substance use that is causing damage to health," either physical or mental.

Because the ICD manual is targeted toward international use, its definition must be broader than the DSM definition intended for use by Americans. Cultural customs of substance use vary widely, sometimes even within the same country.

DEFINITIONS OF DEPENDENCE. In general, the DSM-IV-TR and the ICD-10 manuals agree that dependence is present if three or more of the following occur in a 12-month period:

• Increasing need for more of the substance to achieve the desired effect (tolerance), or a reduction in effect when using the same amount as used previously.

• Withdrawal symptoms if use of the substance is stopped or reduced.

• Progressive neglect of other pleasures and duties.

• A strong desire to take the substance or a persistent but unsuccessful desire to control or reduce the use of the substance.

• Continued use in spite of physical or mental health problems caused by the substance.

• Use of the substance in larger amounts or over longer periods of time than originally intended, or difficulties in controlling the amount of the substance used or when to stop taking it.

• Considerable time spent in obtaining the substance, using it, or recovering from its effects.

Progression from Use to Dependence

The rate at which individuals progress from drug use to drug abuse to drug dependence (or addiction) depends on many of the factors mentioned above. In general, each level is more dangerous, more invasive in the user's life, and more likely to cause social interventions, such as family pressure to enter treatment programs or prison sentences for drug offenses, than the previous level.

FIGURE 1.2

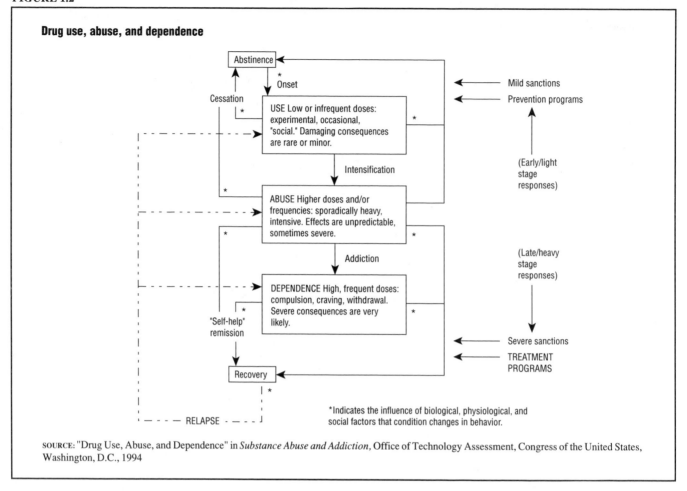

Drug use, abuse, and dependence

SOURCE: "Drug Use, Abuse, and Dependence" in *Substance Abuse and Addiction,* Office of Technology Assessment, Congress of the United States, Washington, D.C., 1994

Figure 1.2 diagrammatically represents these ideas. Notice that the intensification of use leads to abuse, and that abuse leads to addiction and dependence. The right side of the diagram shows social interventions appropriate at various stages of drug use, abuse, and dependence. The dotted lines to the left show that relapse after recovery may lead to renewed drug use, abuse, or dependence.

CHAPTER 2

ALCOHOL, TOBACCO, AND CAFFEINE—CENTURIES OF USE

Alcohol is a drug that affects the brain. For centuries it has been used for medicinal purposes, primarily for sedation. Until 1842, when modern surgical anesthesia began with the use of ether, only heavy doses of alcohol were consistently effective to ease pain during operations.

Tobacco is a commercially grown plant that contains nicotine, an addictive drug. Nicotine is primarily found in tobacco products, such as cigarettes and chewing tobacco, but is also used in the manufacture of certain pesticides.

Caffeine is the most commonly used stimulant in the world. It comes from several commercially grown plants: beans of the coffee plant; leaves of the tea plant; cocoa beans of the cacao tree, from which chocolate is made; and kola nuts from kola trees, from which the flavoring agent in cola drinks is derived. Caffeine is often added to foods and medications. Compared to alcohol and tobacco, it is only mildly addictive, but caffeine can have effects on behavior and health.

Because alcohol, tobacco, and caffeine have been used for so long, they are firmly entrenched in modern life, socially, economically, and politically. Virtually every society uses one or more of these products and must deal with problems they may cause.

ALCOHOL

Ethyl alcohol (ethanol), the active ingredient in beer, wine, and other liquors, is the oldest known mood-altering or psychoactive drug. It is also the only type of alcohol used as a beverage. Ethanol can cause a feeling of well-being or induce sedation, intoxication, or unconsciousness. It can also produce toxic (poisonous) effects on the body. Other alcohols, including methanol and isopropyl alcohol, can have the same toxic effects as ethanol, but much smaller amounts produce severe negative health effects and often death.

Early Fermentation and Distillation

Prehistoric humans probably "discovered" rather than "invented" alcoholic beverages. With the help of airborne yeasts, a fruit or berry mash left in a warm corner of a cave or hut would ferment; that is, the yeast would convert sugars in the fruit to alcohol. Pleased with the effects of these beverages, early humans most likely advanced quickly from accidental discovery to intentional production. Archaeological records of the oldest civilizations indicate the presence of wine and beer.

Until about five hundred years ago, alcoholic beverages were produced only by fermentation and consisted of beers and wines with an alcohol content of up to 14 percent. When this percentage of alcohol is reached, yeasts die and fermentation stops.

In Europe in the fifteenth century, distillation was used to produce alcoholic beverages stronger than fermented wines and beers. These distilled products were known as spirits of wines (usually referred to as "liquors"). Beverages with an alcohol content of 50 percent or more soon became the choice of those desiring quicker or more potent effects.

The distillation process works like this: A liquid is heated to the lowest boiling point of one of the compounds it contains. This compound (such as ethanol) then vaporizes, the vapor is cooled and condensed, and the liquid condensate is collected in a separate container. This process produces a purer and more concentrated liquid. In this case, ethanol is separated from the water and other substances in which it is dissolved, producing a highly concentrated alcoholic product.

The basic characteristics of alcoholic beverages have remained unchanged from early times. Current alcoholic beverages are little more than old recipes refined by technology and produced in much larger quantities.

Early Uses and Abuses

Beer and wine have been used since ancient times in religious rituals, both as a salute to the gods and as sacred drinks from which humans could receive the "divine" power of alcohol. The role of alcohol in religion lent respectability to its later use in secular life. Alcoholic beverages not only were required in worship and the practice of magic and medicine, but also were central to the celebrations of councils, coronations, war, peacemaking, festivals, hospitality, and the rites of birth, initiation, marriage, and death.

In ancient times, just as today, use of beer and wine sometimes led to drunkenness. One of the earliest tracts on temperance was written in Egypt nearly 3,000 years ago. These writings can be thought of as similar to present-day pamphlets espousing moderation in alcohol consumption. Similar recommendations were found in early Greek, Roman, Indian, Japanese, and Chinese writings, as well as in the Bible.

Drinking in America

In colonial America, people drank much more alcohol than they do today, with estimates ranging from three to seven times more alcohol per person per year. Many drank considerable amounts of liquor daily, especially rum, which was readily available through trade with the West Indies.

Liquor was used to ease the pain and discomfort of the common cold, fever, broken limbs, toothaches, frostbite, and just about everything else. Parents often gave liquor to children to relieve their minor aches and pains or to help them sleep. It was also part of many social and religious occasions and was used to give courage to some individuals and to reduce tensions in others. Because of the important role alcohol played, taverns rapidly became the social and political centers of towns.

As early as 1619, drunkenness was illegal in the American colony of Virginia. It was punished in various ways: whipping, placement in the stocks, fines, and even wearing a red "D" (for "drunkard"). By the eighteenth century, all classes of people were getting drunk with greater frequency, even though it was well known that alcohol affected the senses and motor skills, and that drunkenness led to increased crime, violence, accidents, and death.

Temperance

In 1784 Dr. Benjamin Rush, a physician and signer of the Declaration of Independence, published a booklet titled *An Inquiry into the Effects of Ardent Spirits on the Mind and Body*. The pamphlet became popular among the growing number of people concerned about the excessive drinking of many Americans. Such concern gave rise to the temperance movement.

The temperance movement in the United States began in the early 1800s with moderation in drinking as its goal, but by the 1830s, temperance leaders were calling for abstinence from distilled liquors. By the 1850s, large numbers of people were giving up alcohol completely, and by the 1870s, the goal of the temperance movement had become total abstinence from all forms of alcohol.

Middle-class reformers who were concerned about the effects of alcohol on the family, the labor force, and the nation, all of which needed sober participants if they were to remain healthy and productive, generally led the temperance movement. Temperance supporters usually saw alcoholism as a problem of personal morality. They believed that the use of alcohol eventually led the user down the path to ruin. (See Figure 2.1.)

Prohibition

In 1919 reform efforts led to the passage of the Eighteenth Amendment of the U.S. Constitution, which prohibited the "manufacture, sale, or transportation of intoxicating liquors" and their importation and exportation. The Volstead Act of 1919, passed over President Woodrow Wilson's veto, was the Prohibition law that enforced the Eighteenth Amendment.

The Eighteenth Amendment and the Volstead Act did not have the intended effect. Outlawing alcohol did not stop most people from drinking; instead, alcohol was manufactured and sold illegally by gangsters, who organized themselves efficiently and gained considerable political influence from the money they earned. Some of today's organized crime syndicates can be traced back to Prohibition. In addition, the apparent corruption of government and law-enforcement officials contributed to a decline in citizens' respect for these agencies. The Eighteenth Amendment was repealed in 1933 with the passage of the Twenty-first Amendment.

Changing Attitudes

As the decades passed, recognition of the dangers of alcohol increased. In 1956 the American Medical Association endorsed classifying and treating alcoholism as a disease. In 1970 Congress created the National Institute on Alcohol Abuse and Alcoholism, establishing a public commitment to alcohol-related research. During the 1970s, however, many states lowered their drinking age to 18 when the legal voting age was lowered to this age. The rationale was that if people were old enough to vote or to be drafted into the military at 18, they were old enough to drink alcoholic beverages.

Traffic fatalities rose after these laws took effect, and many fatalities involved persons between the ages of 18 and 21 who had been drinking and driving. Organizations such as Mothers Against Drunk Driving (MADD) and Students Against Drunk Driving (SADD) sought to educate the public about the great harm drunk drivers had done to others. As a result, and due to pressure from the federal government, by 1988 all states raised their minimum drinking

FIGURE 2.1

This woodcut (c.1820) illustrates some of the physical and moral perils temperance supporters saw as a result of alcohol use. *(Corbis Corporation (Bellevue).)*

age to 21. Beginning in 1989, warning labels noting the deleterious effects of alcohol on health were required on all retail containers of alcoholic beverages. Courts began to hold restaurants and bars accountable when they permitted obviously intoxicated patrons to drive.

The National Highway Traffic Safety Administration estimates that laws making 21 the minimum drinking age have reduced traffic fatalities involving drivers 18 to 20 years old by 13 percent and have saved an estimated 19,121 lives since 1975 (*Traffic Safety Facts 1999—Alcohol*, National Highway Traffic Safety Administration, 2000). Still, the misuse and abuse of alcohol remain major health and social problems in the world today.

TOBACCO

When Christopher Columbus and his crew arrived in the "New World" in 1492, they found the native inhabitants "perfuming" themselves by puffing on "a lighted firebrand." The Europeans had never seen anyone smoking a pipe.

As European settlers came to North America, American Indians introduced them to tobacco, which is indige-nous to North America and was an important part of American Indian social and religious customs at that time. It was used to communicate with the sacred spirits, to pro-duce visions, and to initiate new shamans (medicine men). Additionally, American Indians believed that tobacco had medicinal properties, so it was used to treat pain, epilepsy, colds, and headaches.

From Pipes to Cigarettes

The use of tobacco, chewed or smoked in pipes, spread quickly throughout Europe during the sixteenth century. In 1560 Jean Nicot (for whom nicotine is named), counselor to the king of France, introduced tobacco to his country. Ben Jonson (1572–1637), an Eng-lish poet and dramatist, said, "Tobacco, I do assert . . . is the most soothing, sovereign and precious weed that ever our dear old mother Earth tendered to the use of man!" Smoking spread as far as Turkey, Russia, and China, although many countries prohibited the use of tobacco.

Despite the disapproval of Louis XIV of France, snuff became fashionable in France during his reign (1643–1715). Snuff is a powdered tobacco that can be chewed, rubbed on the gums, or inhaled through the nose, the process that gave it its name (to snuff means to draw in through the nose).

At about the time of the War of 1812, which was partly funded by tobacco taxes, cigars (small rolls of tobacco leaf for smoking) became fashionable, and were soon more popular than pipes or snuff.

Cigarettes, narrow tubes of cut tobacco enclosed in paper, originated in Brazil during the early 1800s. By the mid-1800s, cigarette smoking was popular in Spain, France, and the United States, although most American tobacco users smoked cigars or chewed tobacco. In the 1880s, however, the cigarette-making machine was invented. It could manufacture 200 cigarettes per minute, or 120,000 in a 10-hour day. Mass production rapidly reduced the price of cigarettes, and sales rose dramatically.

Anti-Smoking Movements

Despite its widespread popularity, tobacco use was not always greeted with enthusiastic approval. James I of England (1603–1625) personally disapproved of tobacco use, forbade tobacco planting in England, and taxed the importation of tobacco. Russia and Turkey outlawed the use of tobacco, imposing penalties of mutilation or even death. A 1683 Chinese law threatened beheading for tobacco users. Frederick the Great of Prussia forbade his mother to use snuff at his 1790 coronation. Louis XV banned snuff from the court of France.

Popes Innocent X and Urban VIII excommunicated smokers from the Roman Catholic Church. Queen Victoria of Great Britain and Ireland (1837–1901) hated the tobacco habit and tried unsuccessfully to outlaw it from the British army. Sylvester Graham (1794–1851), an early health advocate and inventor of the Graham cracker, advised total abstinence from both alcohol and tobacco in order to maintain good health.

Anti-Smoking Efforts in the United States

In the United States, the first anti-smoking movement was organized in the 1830s. Reformers characterized tobacco as a "foul narcotic" and its use as an unhealthy and even fatal habit. Tobacco use was linked to increased alcohol use and lack of cleanliness. Moreover, anti-smoking reformers suggested that tobacco exhausted the soil, wasted money, and promoted laziness, promiscuity, and profanity. It was even blamed for causing baldness and the reading of novels, considered an unwholesome pastime.

The political and social upheaval of the Civil War and its aftermath during the 1860s sidetracked the reform movement for several years, but as the country began to recover from the war, the continued mass production of cigarettes gave the reform movement new life. In 1892 reformers petitioned Congress to prohibit the manufac-

ture, import, and sale of cigarettes. The Senate Committee on Epidemic Diseases agreed that cigarette use was a public health concern but concluded that each state must regulate tobacco matters for itself. By the late 1800s, four states had outlawed cigarette sales to both adults and minors. These bans were later lifted.

World War I again sidetracked the reform movement. The country was prosperous, the troops used cigarettes enthusiastically, and women started to smoke in greater numbers, especially when advertisers associated women's growing freedom with cigarette smoking. All these factors stalled anti-smoking efforts for many years.

In July 1957, following a joint report by the National Cancer Institute, the National Heart Institute, the American Cancer Society, and the American Heart Association, U.S. Surgeon General Leroy E. Burney (a smoker himself) delivered a cautious statement that "the weight of the evidence is increasingly pointing in one direction. . . .that excessive smoking is one of the causative factors in lung cancer." Nevertheless, cigarette ads of the 1950s touted cigarette smoking as pleasurable, sexy, relaxing, flavorful, and/or fun. Later surgeons general imposed stronger warnings and labeling requirements. The "tobacco wars" were under way.

Since 1992, individuals, cities, and states have filed lawsuits against American tobacco companies. In 1998, 46 states, 5 territories, and the District of Columbia signed an agreement with the major tobacco companies to settle all state lawsuits, although the tobacco industry is still liable for class action and individual lawsuits.

CAFFEINE

Stone Age peoples were probably familiar with most of the caffeine-producing plants in existence. Early humans chewed the leaves, bark, and seeds of many plants and learned to enjoy the sensations of alertness and elevated mood produced by some. Consequently, caffeine-producing plants were cultivated widely from early times. It was not until much later, however, that people discovered that steeping the plants in hot water released more of the stimulant. From that discovery came all of our present-day caffeine beverages, including coffee, tea, colas, and cocoa.

Coffee

As early as the sixth century C.E. (Common Era), Ethiopians were cultivating the coffee plant and chewing its berries, although the first written record of coffee was found in tenth-century Arabic documents. At first, the berries were mashed, fermented, and made into a wine called qahwah. It was not until five hundred years later that the Arabians began to brew a hot beverage from roasted coffee beans. They called this beverage qahwah as well, from which the word coffee is derived.

In the 1600s, the Dutch established coffee plantations on Java, an island of Indonesia. By the mid-1700s, the French and British did the same in their Caribbean colonies. Coffee cultivation spread from the Caribbean islands to Central and South America, and by the early 1800s, Brazil was the major producer and exporter of coffee. By the mid-1800s, the United States was the largest consumer of coffee, using more than three-quarters of the world's production of this beverage. At that time, more than half the coffee consumed in the United States was imported from Brazil.

Today, the United States is still the largest consumer of coffee, using annually about one-fifth of all the coffee grown in the world. Other leading coffee consumers are Brazil, France, the United Kingdom, Italy, and Japan. Although Brazil produces about one-fourth of the world's coffee, the crop is vital to the economies of many Latin American countries.

Tea

Tea, called *ch'a* (or *t'e*, pronounced "tay" in the Chinese Amoy dialect), may have been used in China as early as five thousand years ago. Around 600 C.E., many aspects of Chinese culture, including tea drinking, spread to Japan, although tea would not become a regular part of Japanese life for another seven hundred years.

In the seventeenth century, Dutch traders with China introduced tea to Europeans. Although tea was very expensive, its popularity spread quickly throughout Europe, and in some areas tea became more popular than coffee. Tea was particularly popular in the North American colonies, where a visitor in the 1760s reported that American women "would rather go without their dinners than without a dish of tea."

During the colonial period, the British, through their East India Company, had a virtual monopoly on the importation of tea, most of which came from China. The British levied a special tax on tea and other items imported into the American colonies. This tax became a rallying point for colonists dissatisfied with British rule, and Americans began a tea boycott (refusal to buy), primarily using coffee as a substitute. They also destroyed cargoes of tea. On December 16, 1773, a group of citizens disguised themselves as Indians, boarded three ships in Boston Harbor, and dumped the cargoes of tea overboard. This incident, known as the Boston Tea Party, and the reprisals undertaken by the British government helped consolidate resistance to British rule and ultimately hastened the start of the American Revolution.

During the early 1800s, the popularity of tea declined in Britain because of high taxation (the tax on tea was 15 times the domestic tax on coffee). As a result, between 1800 and 1840, coffee use grew tenfold, and coffee became more widely used than tea. A series of coffee adulteration scandals (situations in which contaminants were added to coffee) and reductions in the tea tax led many people to return to tea drinking in the mid-1800s. Today, the countries that consume the most tea per capita include India, Indonesia, Kenya, Sri Lanka, China, Japan, the United Kingdom, Australia, and New Zealand.

Cacao and Kola (Chocolate and Colas)

Both the cacao bean and the kola nut are longtime companions of humans. Cacao (from which we get our word cocoa) is native to Central and South America; as early as 1000 B.C.E. (Before the Common Era), the Mayas, Toltecs, and Aztecs made a drink from roasted cacao beans. On one of his voyages to the New World, Christopher Columbus was served a cacao drink. Since it was unsweetened, however, he found it unpleasantly bitter. Later, in Mexico, Hernando Cortés tasted the chocolate drink of the Aztecs, who sweetened theirs with honey and added spices and vanilla. Cortés liked it so much that he took cacao powder back to Europe with him. Europeans who could afford cacao powder liked this drink as well. Even before coffee and tea were introduced, wealthy Europeans were drinking hot chocolate.

In the early 1800s, the Dutch learned to make cocoa powder from cacao beans. The English learned to make solid dark chocolate and, later, solid milk chocolate. Today, people all over the world enjoy chocolate in hot and cold drinks, cakes, pies, ice cream, and candies.

Kola trees are native to West Africa, where the inhabitants chewed the kola nuts to enjoy the stimulating flavor. The kola nut was first used in a beverage in the United States in the mid-1800s. Although carbonated soft drinks (called that to differentiate them from "hard" alcoholic drinks) were popular in the United States in the early 1800s, they were usually made from local herbs, roots, and other flavorings. A pharmacist in Georgia created the first cola soft drink in the 1880s, making a syrup from coca leaves (from which cocaine is derived), kola nuts, citrus flavoring, cinnamon, and other spices and flavorings. At first, the syrup was mixed with plain water to make the drink, but some enterprising person tried mixing it with carbonated water. The result was the first successful soft drink, Coca-Cola, followed closely by other brands with slightly different formulas.

Many soft drinks on the market today, such as Coca-Cola (and even some orange drinks and root beers), contain caffeine. Americans have more than doubled their consumption of soft drinks since 1970, from 24.3 gallons per person annually to 50.8 gallons per person in 1999. Americans have also increased their annual consumption of tea, from 6.8 gallons per person in 1970 to 8.4 gallons in 1999. Interestingly, however, Americans have reduced their coffee consumption during the same time span. In

1970 per capita consumption of coffee was 33.4 gallons. In 1999 per capita consumption had dropped to 25.7 gallons. In light of these data, the current popularity of coffeehouses may indicate that Americans are drinking more specialty coffees rather than drinking a greater quantity of coffee. Nevertheless, all told, Americans are consuming caffeinated beverages in greater quantities than ever before.

CHAPTER 3

ALCOHOL—WHAT IT IS AND WHAT IT DOES

Contrary to popular belief, ethanol (the alcohol in alcoholic beverages) is not a stimulant; it is a depressant. Although many of those who drink alcoholic beverages feel relaxation, pleasure, and stimulation, these feelings are in fact caused by the depressant effects of alcohol on the brain.

WHAT CONSTITUTES A DRINK?

The typical amount of absolute alcohol in a drink is 0.5 ounces. The following beverages contain approximately equal amounts of alcohol.

- A shot (1.25 ounces) of spirits (80 proof whiskey or vodka—about 40 percent alcohol).

- A 2.5-ounce glass of dessert or cocktail wine (20 percent alcohol).

- A 5-ounce glass of table wine (10 percent alcohol).

- A 12-ounce bottle or can of beer (4.5 percent alcohol).

ALCOHOL CONSUMPTION IN THE UNITED STATES

After caffeine, alcohol is the most commonly used drug in the United States. It is legal, generally acceptable (in moderate amounts), readily available, and relatively inexpensive. Although researchers frequently count how many people are drinking and how often, the statistics do not necessarily reflect the true picture of alcohol consumption in the United States. People tend to underreport their drinking. Furthermore, survey interviewees are typically people living in households. Therefore, the results of survey research may not include the homeless, a portion of the U.S. population traditionally at risk for alcoholism.

Per Capita Consumption of Alcoholic Beverages

According to the Economic Research Service of the U.S. Department of Agriculture, the average per capita consumption of alcoholic beverages peaked at 28.8 gallons in 1981. (The per capita consumption includes the total resident population, including all age groups.) Per capita consumption declined to 25.1 gallons in 1995 and has remained somewhat level since then. In 1999, the average per capita consumption of alcoholic beverages was 25.4 gallons. Table 3.1 shows this trend and also shows the per capita consumption of beer, wine, and distilled spirits for the total population, persons 21 years and older, and those 18 years and older.

Beer remained the most popular alcoholic beverage in 1999, being consumed at a rate of 22.3 gallons per person. Beer consumption peaked in 1981 at 24.6 gallons per person, but its consumption generally declined steadily to 22.0 gallons per person from 1981 to 1995. Since 1995 per capita beer consumption has remained somewhat level. In 1999 the average per capita consumption of beer (22.3 gallons per person) was about the same amount per capita as coffee and milk consumption. The consumption of wine and spirits is much lower; in 1999 consumers drank 1.9 gallons of wine and 1.3 gallons of distilled spirits (liquor) per capita.

A complex set of factors contributes to variations in alcohol use and abuse over individuals' life spans. These factors include psychological and biological mechanisms as well as societal influences. One factor that may contribute to a decrease in alcohol consumption in the future is the changing face of the U.S. population. Those over age 60 generally consume less alcohol than do younger persons, and this older segment of the population is increasing relative to younger segments.

PER CAPITA CONSUMPTION OF ETHANOL. Figure 3.1 shows the total U.S. per capita consumption of ethanol from 1935 to 1998, based on a resident U.S. population of those 14 years of age and older. The data do not distinguish between types of alcoholic beverages but simply show how much alcohol (ethanol) has been consumed per

TABLE 3.1

Per capita consumption of beer, wine, and distilled spirits, 1970–1999

	Total resident population				Resident population, 21 years and older				Resident population, 18 years and older			
	Beer	Wine[2]	Distilled spirits	Total[1]	Beer	Wine[2]	Distilled spirits	Total[1]	Beer	Wine[2]	Distilled spirits	Total[1]
							Gallons					
1970	18.5	1.3	1.8	21.6	30.6	2.2	3.0	35.7	28.1	2.0	2.8	32.8
1971	18.9	1.5	1.8	22.3	31.2	2.4	3.0	36.7	28.6	2.2	2.8	33.6
1972	19.3	1.6	1.9	22.8	31.5	2.6	3.1	37.2	28.8	2.4	2.8	34.1
1973	20.1	1.6	1.9	23.6	32.4	2.7	3.1	38.2	29.7	2.4	2.9	35.0
1974	20.9	1.6	2.0	24.5	33.6	2.6	3.1	39.3	30.7	2.4	2.9	36.0
1975	21.3	1.7	2.0	25.0	33.9	2.7	3.1	39.7	31.0	2.5	2.9	36.3
1976	21.5	1.7	2.0	25.2	33.8	2.7	3.1	39.6	30.9	2.5	2.8	36.2
1977	22.4	1.8	2.0	26.1	34.8	2.8	3.1	40.7	31.8	2.6	2.8	37.2
1978	23.0	2.0	2.0	26.9	35.4	3.0	3.1	41.4	32.4	2.8	2.8	38.0
1979	23.8	2.0	2.0	27.8	36.2	3.0	3.0	42.3	33.3	2.8	2.8	38.8
1980	24.3	2.1	2.0	28.3	36.6	3.2	3.0	42.8	33.7	2.9	2.7	39.4
1981	24.6	2.2	2.0	28.8	36.9	3.3	2.9	43.1	34.0	3.0	2.7	39.7
1982	24.4	2.2	1.9	28.5	36.3	3.3	2.8	42.3	33.5	3.0	2.6	39.1
1983	24.2	2.3	1.8	28.3	35.7	3.3	2.7	41.8	33.1	3.1	2.5	38.7
1984	24.0	2.4	1.8	28.1	35.1	3.4	2.6	41.2	32.6	3.2	2.5	38.3
1985	23.8	2.4	1.8	28.0	34.6	3.5	2.6	40.7	32.3	3.3	2.4	38.0
1986	24.1	2.4	1.6	28.2	34.9	3.5	2.4	40.8	32.6	3.3	2.2	38.2
1987	24.0	2.4	1.6	28.0	34.6	3.5	2.3	40.4	32.4	3.2	2.2	37.8
1988	23.8	2.3	1.5	27.6	34.3	3.2	2.2	39.8	32.1	3.0	2.1	37.2
1989	23.6	2.1	1.5	27.2	33.9	3.1	2.2	39.1	31.7	2.9	2.0	36.6
1990	23.9	2.0	1.5	27.5	34.4	2.9	2.2	39.5	32.2	2.7	2.0	37.0
1991	23.1	1.8	1.4	26.4	33.2	2.7	2.0	37.8	31.2	2.5	1.9	35.6
1992	22.8	1.9	1.4	26.1	32.7	2.7	2.0	37.3	30.8	2.5	1.9	35.2
1993	22.6	1.7	1.3	25.7	32.4	2.5	1.9	36.8	30.6	2.4	1.8	34.7
1994	22.5	1.8	1.3	25.6	32.2	2.5	1.8	36.6	30.5	2.4	1.7	34.6
1995	22.0	1.8	1.2	25.1	31.6	2.6	1.8	35.9	29.8	2.4	1.7	33.9
1996	22.1	1.9	1.2	25.2	31.6	2.7	1.8	36.1	29.8	2.6	1.7	34.1
1997	22.0	2.0	1.2	25.2	31.5	2.8	1.8	36.1	29.7	2.6	1.7	34.0
1998	22.1	1.9	1.2	25.2	31.7	2.6	1.8	36.1	29.8	2.5	1.7	34.0
1999	22.3	1.9	1.3	25.4	31.9	2.7	1.8	36.3	30.0	2.5	1.7	34.2

[1] Computed from unrounded data.
[2] Beginning in 1983, includes wine coolers.

SOURCE: "Beverages: Per capita consumption, 1970–99" USDA/Economic Research Service

year based on total sales and/or shipments. This graph shows two peaks of alcohol consumption: one is during the years of American involvement in World War II (1941–45) and its aftermath. Alcohol consumption then declined through the late 1940s and the 1950s, but increased dramatically in the 1960s and 1970s, peaking in 1980–81 at 2.76 gallons per person. Since 1981 alcohol consumption has generally declined. In 1998 the U.S. per capita consumption was 2.20 gallons of ethanol.

How Many Americans Drink Alcohol?

The Substance Abuse and Mental Health Services Administration (SAMHSA), in the *National Household Survey on Drug Abuse: Population Estimates 1998* (Rockville, Maryland, 1999), reported that about 81.3 percent of the U.S. population (or about 177.5 million Americans) 12 years of age and over had tried an alcoholic beverage at least once in their lives. Nearly two-fifths (37.3 percent) of those 12 to 17 years old had tried alcohol at some time in their lives, as had an estimated 83.2 percent of those between the ages of 18 and 25 years. About 64.0 percent of Americans (139.8 million) had con-

sumed at least one drink during the year prior to the survey, while 51.7 percent (112.8 million) had used alcohol in the month prior to the survey. (See Table 3.2.)

In all age categories shown in Table 3.2, except in the "Ever Used" or "Used Past Year" categories for those aged 12 to 17 years, men or boys were more likely to drink alcoholic beverages than were women or girls. However, in this age group as well as in all others, men and boys were more likely than women and girls to have used alcohol during the past month. In comparing the years 1997 and 1998, Table 3.3 also shows that a higher percentage of males reported using alcohol during the past month than did females. Data in this table also show that adult alcohol use in both 1997 and 1998 was highest among those aged 18 to 34, and over half of men and nearly half of women aged 35 years and older used alcohol in 1997 and 1998. Other statistics to note in this table are that across age groups, a higher percentage of whites had used alcohol within the month prior to the survey than had blacks or Hispanics, and past month alcohol use increased with level of education.

FIGURE 3.1

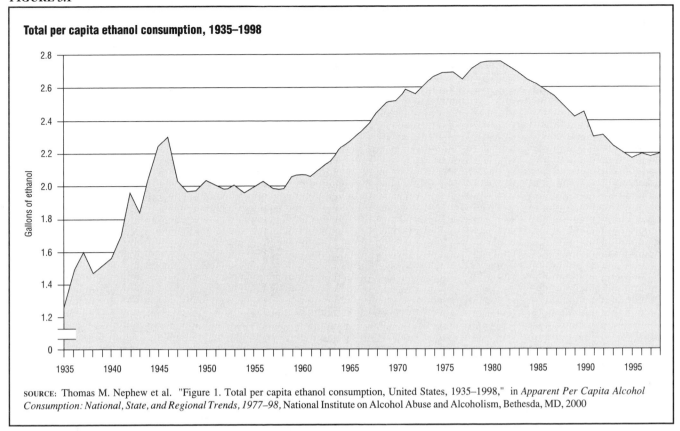

Total per capita ethanol consumption, 1935–1998

SOURCE: Thomas M. Nephew et al. "Figure 1. Total per capita ethanol consumption, United States, 1935–1998," in *Apparent Per Capita Alcohol Consumption: National, State, and Regional Trends, 1977–98,* National Institute on Alcohol Abuse and Alcoholism, Bethesda, MD, 2000

National Health Interview Survey—Excessive Alcohol Consumption

The National Health Interview Survey (NHIS) is one of the major data collection programs of the National Center for Health Statistics (NCHS). The NCHS is one of 12 centers, institutes, and offices of the Centers for Disease Control and Prevention (CDC) located in Atlanta, Georgia. The CDC, an agency of the United States Department of Health and Human Services (HHS), is the lead federal agency charged with protecting the health and safety of the people of the United States.

The National Health Interview Survey has been conducted annually since 1957. Figure 3.2 shows recent NHIS data on excessive alcohol consumption. The 2000 NHIS estimate of the percentage of adults with excessive alcohol consumption was 8.8 percent. This estimate is not statistically different from the estimates for 1997–99, which are shown on the graph. This means that approximately the same percentage of people consumed alcohol excessively each year from 1997–2000. The variation shown from year to year can be due simply to chance variations within the samples of the population from which the data were taken. The brackets at the top of the bars show the amount of variation likely for each year.

The Elderly

A number of surveys conducted in health care settings have found an increasing prevalence of alcoholism among the older population. The prevalence of problem drinking in nursing homes is as high as 49 percent in some studies. It appears that some people increase their alcohol consumption later in life, occasionally leading to late-onset alcoholism. In general, people who are not heavy drinkers until later in life are in better physical, mental, and social health than those who started drinking heavily at an earlier age. They also respond better to intervention and treatment.

The National Health and Nutrition Examination Survey (NHANES) is conducted by the National Center for Health Statistics. This survey collects information about the health and diet of people in the United States. NHANES combines a home interview with health tests, which are done in a Mobile Examination Center.

Dr. Alison A. Moore, et al., in "Drinking Habits Among Older Persons: Findings from the NHANES I Epidemiologic Followup Study (1982–84)" (*Journal of the American Geriatrics Society*, April 1999), found that 60 percent of the people 65 years of age and older who took part in NHANES I regularly consumed alcohol at some time in their lives. Of these people, 79 percent were still regular drinkers, one-fourth of whom drank daily. Sixteen percent of the men and 15 percent of the women were heavy drinkers (two or more drinks per day for men and more than one drink a day for women). Ten percent of people over age 65 could be classified as binge drinkers, those who consume more than five drinks in succession.

TABLE 3.2

Alcohol use by gender within age group for total population, 1998

AGE/GENDER	Ever Used		Used Past Year		Used Past Month	
	Observed Estimate	95% C.I.	Observed Estimate	95% C.I.	Observed Estimate	95% C.I.
	RATE ESTIMATES (Percent)					
12-17	37.3	(35.5 - 39.2)	31.8	(30.0 - 33.7)	19.1	(17.5 - 20.7)
Male	36.6	(34.2 - 39.1)	31.0	(28.6 - 33.4)	19.4	(17.5 - 21.5)
Female	38.1	(35.7 - 40.7)	32.7	(30.3 - 35.1)	18.7	(16.7 - 21.0)
18-25	83.2	(81.4 - 84.8)	74.2	(71.8 - 76.4)	60.0	(57.3 - 62.7)
Male	87.2	(85.2 - 88.9)	79.3	(76.5 - 81.9)	68.2	(64.9 - 71.2)
Female	79.2	(76.4 - 81.7)	68.9	(65.6 - 72.1)	51.7	(48.0 - 55.5)
26-34	88.2	(86.7 - 89.5)	74.5	(72.5 - 76.4)	60.9	(58.7 - 63.0)
Male	90.9	(89.0 - 92.4)	77.7	(74.9 - 80.2)	67.7	(64.3 - 70.9)
Female	85.5	(83.4 - 87.4)	71.5	(68.5 - 74.2)	54.2	(51.4 - 56.9)
≥35	86.6	(85.1 - 87.9)	64.6	(62.6 - 66.6)	53.1	(51.0 - 55.1)
Male	92.2	(90.6 - 93.5)	70.2	(67.7 - 72.6)	61.4	(58.7 - 64.0)
Female	81.6	(79.5 - 83.5)	59.7	(57.0 - 62.3)	45.8	(43.2 - 48.4)
TOTAL	81.3	(80.1 - 82.4)	64.0	(62.4 - 65.6)	51.7	(50.0 - 53.3)
Male	85.2	(83.9 - 86.3)	68.3	(66.5 - 70.1)	58.7	(56.7 - 60.7)
Female	77.6	(76.0 - 79.2)	60.0	(57.9 - 62.0)	45.1	(43.2 - 47.1)
	POPULATION ESTIMATES (In Thousands)					
12-17	8,491	(8,075 - 8,916)	7,233	(6,819 - 7,661)	4,338	(3,990 - 4,709)
Male	4,246	(3,965 - 4,536)	3,596	(3,324 - 3,880)	2,255	(2,029 - 2,500)
Female	4,245	(3,969 - 4,528)	3,637	(3,372 - 3,912)	2,083	(1,854 - 2,333)
18-25	23,268	(22,778 - 23,720)	20,739	(20,082 - 21,359)	16,786	(16,019 - 17,535)
Male	12,313	(12,036 - 12,557)	11,203	(10,803 - 11,566)	9,628	(9,170 - 10,063)
Female	10,955	(10,572 - 11,304)	9,536	(9,075 - 9,972)	7,158	(6,638 - 7,674)
26-34	30,506	(29,998 - 30,964)	25,794	(25,102 - 26,452)	21,069	(20,313 - 21,810)
Male	15,600	(15,278 - 15,872)	13,334	(12,861 - 13,769)	11,624	(11,045 - 12,174)
Female	14,906	(14,536 - 15,235)	12,460	(11,950 - 12,940)	9,444	(8,959 - 9,925)
≥35	115,247	(113,355 - 116,984)	86,041	(83,368 - 88,646)	70,657	(67,913 - 73,387)
Male	57,425	(56,466 - 58,238)	43,758	(42,189 - 45,254)	38,236	(36,548 - 39,881)
Female	57,822	(56,353 - 59,173)	42,283	(40,385 - 44,140)	32,421	(30,572 - 34,287)
TOTAL	177,512	(174,928 - 179,975)	139,807	(136,301 - 143,246)	112,850	(109,261 - 116,431)
Male	89,584	(88,263 - 90,819)	71,891	(69,985 - 73,740)	61,744	(59,617 - 63,840)
Female	87,928	(86,029 - 89,730)	67,916	(65,580 - 70,212)	51,107	(48,899 - 53,332)

SOURCE: "Table 13A: Alcohol Use by Gender Within Age Group for Total Population in 1998," in *National Household Survey on Drug Abuse: Population Estimates 1998,* Substance Abuse and Mental Health Services Administration, Rockville, MD, 1999

WHAT ALCOHOL DOES IN THE BODY

When most people think about how alcohol affects them, they think of a temporary light-headedness or a hangover the next morning. Many are also aware of the serious damage that continuous, excessive alcohol use can do to the liver. Alcohol, however, affects many organs of the body and has been linked to cancer, mental and/or physical retardation in newborns, heart disease, and other health problems.

While most of the effects from alcohol occur after it enters the bloodstream, the lining of the gastrointestinal tract (stomach and intestines) is also affected. Low concentrations of alcohol cause the secretion of gastric (stomach) juices. This is why a glass of wine before a meal is said to stimulate the appetite. A high concentration of alcohol, however, irritates the lining of the stomach and intestines.

Low to moderate doses of alcohol produce a slight, brief increase in heartbeat and blood pressure. Large doses can reduce the pumping power of the heart producing irregular electrocardiograms (EKGs), which are recordings of the electrical activity of the heart. Blood vessels within muscles constrict, but those at the surface expand, causing rapid heat loss from the skin. This causes the flushing or reddening of the skin that often accompanies alcohol intake. Large doses of alcohol decrease body temperature but may cause numbness of the skin, legs, and arms, creating a false feeling of warmth. Figure 3.3 illustrates the path alcohol takes through the body after it is consumed.

Alcohol affects the endocrine system (a group of glands that produce hormones) in several ways. One effect is increased urination. Urination increases not only because of fluid intake, but also because alcohol stops the release of an antidiuretic hormone—ADH, or vasopressin—from the pituitary gland. This hormone controls how much water the kidneys reabsorb from the urine as it is being made, and how much they excrete. Therefore, heavy alcohol intake can result in both dehydration and an imbalance in electrolytes, which are chemicals dissolved in body fluids that conduct electrical currents. Both of these conditions are serious health hazards.

TABLE 3.3

Percentages reporting past month use of alcohol, by age group and demographic characteristics, 1997 and 1998

| Demographic Characteristic | AGE GROUP (Years) | | | | | | | | | |
| | 12-17 | | 18-25 | | 26-34 | | 35 and Older | | Total | |
	1997	1998	1997	1998	1997	1998	1997	1998	1997	1998
TOTAL	20.5	19.1	58.4	60.0	60.2	60.9	52.8	53.1	51.4	51.7
RACE/ ETHNICITY										
White, non-Hispanic	22.0	20.9	63.5	65.0	64.8	65.2	56.1	56.2	55.1	55.3
Black, non-Hispanic	16.3	13.1	46.6	50.3	51.0	54.8	40.9	38.3	40.4	39.8
Hispanic	18.8	18.9	48.5	50.8	51.6	53.1	42.8	47.7	42.4	45.4
Other, non-Hispanic	17.0	12.5	*	45.5	42.2	39.1	36.0	37.5	37.0	35.8
GENDER										
Male	21.0	19.4	65.9	68.2	67.9	67.7	60.6	61.4	58.2	58.7
Female	19.9	18.7	50.8	51.7	52.6	54.2	46.0	45.8	45.1	45.1
POPULATION DENSITY[1]										
Large Metro	21.4[a]	17.7	58.5	58.2	62.4	61.2	56.6	57.0	54.3	53.9
Small Metro	20.4	20.0	58.7	63.1	62.3	61.5	53.5	54.6	52.0	53.3
Nonmetro	19.0	20.1	58.0	57.4	51.0[a]	59.0	45.4	43.4	44.9	44.7
REGION										
Northeast	19.3	22.1	59.9	64.3	60.3	62.4	56.7	57.7	54.0	55.8
North Central	23.4	20.6	68.0	72.1	66.6	67.9	55.3	58.7	55.6	57.8
South	19.3	17.5	55.1	54.4	57.9	58.3	47.7	45.6	47.4	45.8
West	20.3	17.4	53.3	53.0	56.8	56.5	55.7	55.0	51.7	50.9
ADULT EDUCATION[2]										
<High School	N/A	N/A	52.4	50.8	48.7	51.5	32.7	36.0	38.0	40.4
High School Grad	N/A	N/A	56.7	57.6	56.9	59.0	52.7	49.4	54.0	52.3
Some College	N/A	N/A	61.3	64.6	61.6	64.0	56.0	57.3	58.2	60.1
College Graduate	N/A	N/A	67.4	70.5	68.9	64.3	65.9	65.5	66.6	65.5
CURRENT EMPLOYMENT[2]										
Full-time	N/A	N/A	65.3	66.6	64.0	64.4	60.4	60.9	61.9	62.5
Part-time	N/A	N/A	51.5	56.0	59.9	59.9	58.4	59.1	56.6	58.4
Unemployed	N/A	N/A	55.8	63.3	55.7	54.5	61.7	59.8	58.8	59.7
Other[3]	N/A	N/A	51.3	49.4	44.6	45.3	40.8	40.6	42.1	41.9

*Low precision; no estimate reported.
N/A: Not applicable.
[1] Population density is based on 1990 MSA classifications and their 1990 Census of Population counts.
[2] Data on adult education and current employment not shown for persons aged 12–17. Estimates for both adult education and current employment are for persons aged ≥ 18.
[3] Retired, disabled, homemaker, student, or "other."
[a] Difference between 1997 and 1998 is statistically significant at the .05 level.

SOURCE: "Table 22: Percentages Reporting Past Month Use of Alcohol, by Age Group and Demographic Characteristics: 1997 and 1998" in *Summary of Findings from the 1998 National Household Survey on Drug Abuse,* Substance Abuse and Mental Health Services Administration, Rockville, MD, 1999

Alcohol also inhibits the release of the hormone oxytocin from the pituitary gland. Oxytocin normally stimulates contractions of the uterus during childbirth. Because alcohol inhibits uterine contractions, it was used in the past to control premature labor.

Alcohol is sometimes believed to be an aphrodisiac (sexual stimulant). While low to moderate amounts of alcohol can reduce fear and decrease sexual inhibitions, larger doses tend to impair sexual performance. Alcoholics often reveal that their sex lives are disturbed, deficient, and ineffectual.

Intoxication

The speed of alcohol absorption affects the rate at which one becomes intoxicated. Intoxication occurs when alcohol is absorbed into the blood faster than the liver can oxidize it (or break it down into water, carbon dioxide, and energy). In a 160-pound man, alcohol is metabolized (absorbed and processed by the body) at a rate of about one drink every two hours. Heavier people are less affected than lighter people by the same amount of alcohol because there is more blood and water in their systems to dilute the alcohol intake. The absorption of alcohol is influenced by several factors:

• Body weight—The greater the body muscle (not fat) weight, the lower the blood alcohol concentration (BAC) for a given amount of alcohol.

• Speed of drinking—The faster alcohol is drunk, the faster the BAC level rises. Not all the alcohol consumed is metabolized; some of it remains in the bloodstream until it is eliminated as waste.

FIGURE 3.2

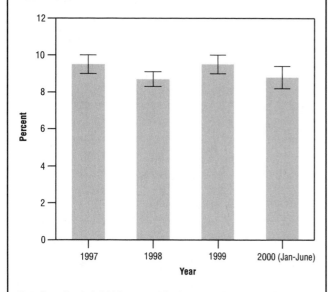

Percentage of adults with excessive alcohol consumption, 1997–2000

Note: Excessive alcohol drinkers were defined as those who had ≥12 drinks of any type of alcoholic beverage in their lifetime *and* consumed ≥5 drinks on one occasion for at least 12 times during the year preceeding the interview. The analysis excluded adults with unknown alcohol consumption. Brackets indicate 95% confidence intervals (CI).

SOURCE: "Percentage of adults with excessive alcohol consumption: United States, 1997–2000," in *National Health Interview Survey, 2000,* Centers for Disease Control and Prevention, Atlanta, 2000

As a person's BAC rises, there are somewhat predictable responses in behavior.

- At a BAC of about 0.05 percent (0.05 grams of alcohol per 100 milliliters of blood), thought processes, judgment, and restraint are more lax. The person may feel more at ease socially. Also, reaction time to visual or auditory stimuli slows down as the BAC rises.

- At 0.10 percent, the level in most states at which it is illegal to drive, voluntary motor actions become noticeably clumsy. (Nearly half the states and the District of Columbia have lowered the legal limit to 0.08 percent.)

- At 0.20 percent, the entire motor area of the brain becomes significantly depressed. The person staggers, may want to lie down, may be easily angered, or may shout or weep.

- At 0.30 percent, the person generally acts confused or may be in a stupor.

- At 0.40 percent, the person usually falls into a coma.

- At 0.50 percent or more, the medulla area of the brain is severely depressed, and death generally occurs within a couple of hours, usually from respiratory failure.

There have been some cases of delayed death from circulatory failure as much as 16 hours after the last known ingestion of alcohol. A BAC of 0.50 percent without immediate medical attention is almost always fatal. Death can occur even at 0.40 percent if the alcohol is "chugged" in a large amount, causing the BAC to rise rapidly.

Sobering Up

Time is the only way to rid the body of alcohol. The more slowly a person drinks, the more time the body has to process the alcohol, so that less alcohol accumulates in the bloodstream. According to the National Clearinghouse for Alcohol and Drug Information, five drinks consumed close together by a 180-pound man will produce a BAC of 0.11. In a 140-pound man, this intake will produce a BAC of 0.13. And in a 120-pound woman, it will produce a BAC of 0.19. The body takes nearly seven hours to metabolize this blood concentration of alcohol. Under normal conditions, five drinks consumed with an hour or so between each drink will produce a BAC of only 0.02, depending on the gender and weight of the person. It will likely produce a BAC higher than 0.02 in women and persons weighing less than 180 pounds.

Hangovers

Hangovers cause a great deal of misery as well as absenteeism and loss of productivity at work or school. The major symptoms of the aftermath of drinking—headache, dizziness, nausea, thirst, fatigue, and irritability—are well known (see Table 3.4), but the causes of these symptoms are less well known. Some researchers

- Presence of food in the stomach—Eating while drinking slows down the absorption of alcohol by increasing the amount of time it takes the alcohol to get from the stomach to the small intestine.

- Drinking history and body chemistry—The longer a person has been drinking, the greater is his or her tolerance (i.e., the more alcohol it takes him or her to get "high"). An individual's physiological functioning or "body chemistry" may also affect his or her reactions to alcohol. Women are more easily affected by alcohol regardless of weight because they have less of the metabolizing enzyme that processes alcohol in their stomachs, which permits more of it to enter the bloodstream.

Increasing amounts of alcohol affect different areas of the brain. One or two drinks affect the surface of the brain. Three to four drinks reach the cerebrum, which regulates conscious control and reasoning. Five to seven drinks alter the medulla's ability to control involuntary responses such as heartbeat, breathing, and flow of digestive juices. Eight to twelve drinks can influence the cerebellum, which controls the skeletal muscles and governs the body's ability to maintain balance. (See Figure 3.4. Blood alcohol level, or BAL, in this figure is the same as blood alcohol concentration, or BAC.)

FIGURE 3.3

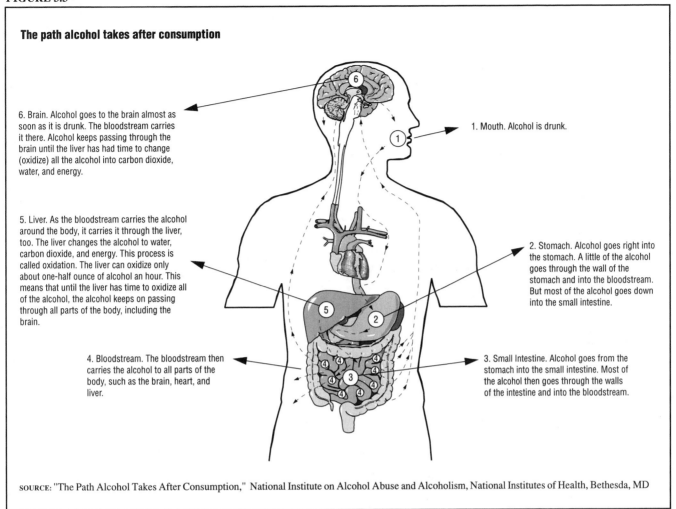

The path alcohol takes after consumption

6. Brain. Alcohol goes to the brain almost as soon as it is drunk. The bloodstream carries it there. Alcohol keeps passing through the brain until the liver has had time to change (oxidize) all the alcohol into carbon dioxide, water, and energy.

5. Liver. As the bloodstream carries the alcohol around the body, it carries it through the liver, too. The liver changes the alcohol to water, carbon dioxide, and energy. This process is called oxidation. The liver can oxidize only about one-half ounce of alcohol an hour. This means that until the liver has time to oxidize all of the alcohol, the alcohol keeps on passing through all parts of the body, including the brain.

4. Bloodstream. The bloodstream then carries the alcohol to all parts of the body, such as the brain, heart, and liver.

1. Mouth. Alcohol is drunk.

2. Stomach. Alcohol goes right into the stomach. A little of the alcohol goes through the wall of the stomach and into the bloodstream. But most of the alcohol goes down into the small intestine.

3. Small Intestine. Alcohol goes from the stomach into the small intestine. Most of the alcohol then goes through the walls of the intestine and into the bloodstream.

SOURCE: "The Path Alcohol Takes After Consumption," National Institute on Alcohol Abuse and Alcoholism, National Institutes of Health, Bethesda, MD

suspect low blood sugar as a cause, while others think that fluid retention in the brain, producing swelling of the cerebrum, may be responsible.

There is no scientific evidence to support popular hangover cures such as black coffee, raw egg, chili pepper, steak sauce, "alkalizers," and vitamins. Health care practitioners usually prescribe bed rest as well as eating and an increased fluid intake.

HEALTH CONSEQUENCES

Research by J. Michael Gaziano and colleagues, ("Light-to-Moderate Alcohol Consumption and Mortality in the Physicians' Health Study Enrollment Cohort," *Journal of the American College of Cardiology*, 2000), shows that light-to-moderate drinkers (one to nearly two drinks per day) have a lowered risk of cardiovascular disease (CVD). These results are similar to results found in earlier studies. At two or more drinks per day, however, the CVD benefits begin to be offset by an increasing risk of some of the less common cancers, such as cancer of the stomach, throat, or esophagus, which have been previously associated with alcohol use.

Researchers speculate that a light-to-moderate intake of alcohol increases the concentration of high density lipoprotein (HDL) in the blood, the "good" lipoprotein that protects against cholesterol build-up in the arteries. HDL may also modify the clotting activity of the blood and affect coronary artery diameter and coronary blood flow. In addition, moderate alcohol consumption may lower stroke risk, perhaps reducing the tendency of blood platelets to clump and form blood clots.

Until menopause, women have a significantly lower death rate from heart disease than men, due to the female hormone estrogen. After menopause, the rate of heart disease in women increases significantly. Studies have shown that in postmenopausal women, three to six drinks per week may reduce the risk of cardiovascular disease without significantly impairing bone density or increasing the risk of alcoholic liver disease. However, alcohol use is linked to increased risk of developing breast cancer. Having one alcoholic drink per day increases a woman's risk of breast cancer slightly. Two to five drinks per day increase breast cancer risk about one and one-half times more than in women who do not drink alcohol.

FIGURE 3.4

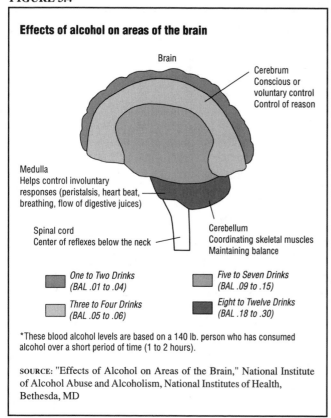

Effects of alcohol on areas of the brain

Brain

Cerebrum
Conscious or
voluntary control
Control of reason

Medulla
Helps control involuntary
responses (peristalsis, heart beat,
breathing, flow of digestive juices)

Spinal cord
Center of reflexes below the neck

Cerebellum
Coordinating skeletal muscles
Maintaining balance

One to Two Drinks
(BAL .01 to .04)

Five to Seven Drinks
(BAL .09 to .15)

Three to Four Drinks
(BAL .05 to .06)

Eight to Twelve Drinks
(BAL .18 to .30)

*These blood alcohol levels are based on a 140 lb. person who has consumed
alcohol over a short period of time (1 to 2 hours).

SOURCE: "Effects of Alcohol on Areas of the Brain," National Institute
of Alcohol Abuse and Alcoholism, National Institutes of Health,
Bethesda, MD

TABLE 3.4

Symptoms of hangover

Class of Symptoms	Type
Constitutional	Fatigue, weakness, and thirst
Pain	Headache and muscle aches
Gastrointestinal	Nausea, vomiting, and stomach pain
Sleep and biological rhythms	Decreased sleep, decreased REM,[1] and increased slow-wave sleep
Sensory	Vertigo and sensitivity to light and sound
Cognitive	Decreased attention and concentration
Mood	Depression, anxiety, and irritability
Sympathetic hyperactivity	Tremor, sweating, and increased pulse and systolic blood pressure

[1] REM = rapid eye movements

SOURCE: Robert Swift and Dena Davidson, "Symptoms of Hangover," in
"Alcohol Hangover," in *Alcohol Health & Research World,* Vol 22,
No. 1, 1998

FIGURE 3.5

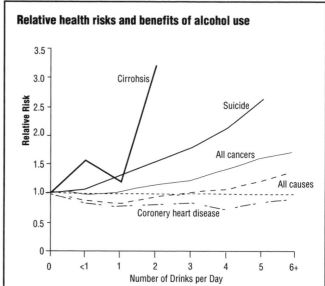

Relative health risks and benefits of alcohol use

Relative risk of dying from various causes for middle-aged men consuming
different alcohol amounts. More than 270,000 Caucasian men ages 40 to 59 were
followed for 12 years; their death rates and causes of death were analyzed
according to their alcohol-consumption levels. The relative risk is the ratio of the
death rate from a specific cause among a certain group of drinkers to the death rate
from the same cause among abstainers. A relative risk of less than 1.0 indicates a
protective effect of alcohol consumption; a relative risk of greater than 1.0 indicates
a detrimental alcohol effect.

SOURCE: Mary C. Dufour, "Relative health risks of alcohol use" in
"Risks and Benefits of Alcohol Use over the Life Span," *Alcohol Health
& Research World,* vol. 20, no. 3, 1996

Results of a study by Morten Gronbaek and his colleagues, "Population-based Cohort Study of the Association between Alcohol Intake and Cancer of the Upper Digestive Tract" (*British Medical Journal*, September 1998), show that intake of wine tends to decrease the risk of upper digestive tract cancer, while intake of beer and spirits significantly increases the risk. The researchers suggest that their findings are strongly supported by studies showing that resveratrol, a substance in grapes and wine, may be protective against cancer, while nitrosamines in beer and spirits may promote cancer.

Although drinking alcoholic beverages has positive effects on cardiovascular health, Figure 3.5 shows that risk of death from all causes increases above the risk for nondrinkers at about three drinks per day. The risk of death from all cancers increases at about one and one-half drinks per day. Suicide risk increases with any amount of alcoholic beverages and increases sharply as the number of drinks per day increases. The risk for developing cirrhosis of the liver also increases with any amount of drinking alcoholic beverages. Cirrhosis of the liver is a disease in which alcohol kills liver cells, which are replaced by connective tissue. Liver cirrhosis is a serious disease and can lead to death.

Alcohol abuse and dependence cause major public-health problems. In 1995, according to the National Institute on Alcohol Abuse and Alcoholism (NIAAA), national costs related to alcohol were estimated to be $166.5 billion. Alcohol is also estimated to be a contributing factor to or direct cause of nearly 111,000 deaths each year. According to the NIAAA, slightly over 18 million Americans age 18 and over could be classified as either alcohol abusers or alcoholics during the year 2000.

Alcohol-Related Hospitalization

According to the NIAAA, about 421,000 hospital visits in 1997 resulted in first-listed (primary) alcohol-related

TABLE 3.5

Definition of alcohol-related diagnoses

Category Used in Report	Classification in ICD-9-CM	
Alcoholic psychoses	291.0	Alcohol withdrawal delirium
	291.1	Alcohol amnestic syndrome
	291.2	Other alcoholic dementia
	291.3	Alcohol withdrawal hallucinosis
	291.4	Idiosyncratic alcohol intoxication
	291.5	Alcoholic jealousy
	291.8	Other specified alcoholic psychosis
	291.9	Unspecified alcoholic psychosis
Alcohol dependence syndrome	303.0	Acute alcoholic intoxication
	303.9	Other and unspecified alcohol dependence
	265.2	Pellagra
	357.5	Alcoholic polyneuropathy
	425.5	Alcoholic cardiomyopathy
	535.3	Alcoholic gastritis
Nondependent abuse of alcohol	305.0	Alcohol abuse
Chronic liver disease and cirrhosis:		
Alcoholic cirrhosis of the liver	571.0	Alcoholic fatty liver
	571.1	Acute alcoholic hepatitis
	571.2	Alcoholic cirrhosis of liver
	571.3	Alcoholic liver damage, unspecified
Other specified cirrhosis of the liver without mention of alcohol	571.4	Chronic hepatitis
	571.6	Biliary cirrhosis
	571.8	Other chronic nonalcoholic liver disease
	572.3	Portal hypertension
Unspecified cirrhosis of the liver without mention of alcohol	571.5	Cirrhosis of liver without mention of alcohol
	571.9	Unspecified chronic liver disease without mention of alcohol

SOURCE: Christine C. Whitmore, Frederick S. Stinson, and Mary C. Dufour, "Definition of Alcohol-Related Diagnoses," in *Trends in Alcohol-Related Morbidity Among Short-Stay Community Hospital Discharges, United States, 1979–97,* National Institute on Alcohol Abuse and Alcoholism, Bethesda, MD, 1999

FIGURE 3.6

Percent distribution of first-listed alcohol-related diagnosis, 1997

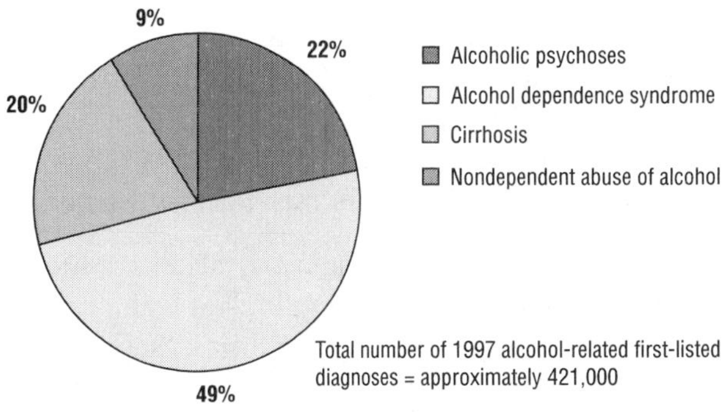

■ Alcoholic psychoses
□ Alcohol dependence syndrome
▨ Cirrhosis
▨ Nondependent abuse of alcohol

Total number of 1997 alcohol-related first-listed diagnoses = approximately 421,000

SOURCE: Christine C. Whitmore, Frederick S. Stinson, and Mary C. Dufour, "Figure 1. Percent distribution of first-listed diagnosis among discharges with first-listed mention of an alcohol-related diagnosis, 1997," in *Trends in Alcohol-Related Morbidity Among Short-Stay Community Hospital Discharges, United States, 1979–97,* National Institute on Alcohol Abuse and Alcoholism, Bethesda, MD, 1999

diagnoses. Alcohol was also listed as a contributing factor in another one million hospital visits. Table 3.5 lists the alcohol-related diagnostic categories.

Alcohol dependence syndrome accounted for 49 percent of first-listed diagnoses related to alcohol. Alcoholic psychosis made up 22 percent; cirrhosis, 20 percent; and nondependent abuse of alcohol, 9 percent (see Figure 3.6).

The length of hospital stays varies by diagnosis. In 1997 the hospital stay of persons diagnosed as alcohol dependent averaged 6.3 days, while that of those with cirrhosis averaged 7.2 days. Those diagnosed with alcoholic psychoses stayed an average of 5.7 days, while those with nondependent abuse of alcohol were in the hospital for 4.5 days.

Liver Diseases

Because the liver plays a central role in removing substances from the body, it is the major organ damaged by chronic drinking. This damage is called alcoholic liver disease (ALD) and occurs in three categories: fatty liver, hepatitis, and cirrhosis.

Alcohol consumption produces changes in the metabolism of lipids (fats) that can cause these compounds to accumulate in liver cells, producing "fatty liver." In 1998 the NIAAA reported that fatty liver was the most common form of ALD. Often there are no external signs of this ailment except in severe cases. This condition is usually reversible and improves with abstinence from alcohol.

Hepatitis, which means "inflammation of the liver," can develop if heavy drinking continues. The liver becomes enlarged and tender, and jaundice (yellowing of the skin and other tissues) is usually present. The condition can be, but is not always, fatal. Some complications, however, may cause hepatitis to persist for long periods after the person stops drinking. The NIAAA reported in 1998 that approximately 10 to 35 percent of heavy drinkers develop alcoholic hepatitis, which is often a precursor of cirrhosis. Women have a higher incidence of alcoholic hepatitis than do men.

Cirrhosis of the liver is the most advanced stage of alcoholic liver disease. It is an inflammatory condition in which functioning liver cells are replaced by scar tissue. While this condition usually follows fatty liver and hepatitis, it can occur without any previous liver disease. If cirrhosis is allowed to persist and further drinking continues, the disease is fatal.

Not all alcoholics develop cirrhosis of the liver. Steady, heavy drinking (five or more drinks a day for several years) is necessary to produce enough liver damage to cause cirrhosis. An estimated 10 to 20 percent of heavy drinkers develop alcoholic cirrhosis. Long-term, heavy drinking, however, is the single most prevalent cause of cirrhosis. In 1998 medical experts estimated that between

10,000 and 24,000 deaths due to cirrhosis could be attributed to excessive alcohol use.

Women develop cirrhosis more easily with less alcohol consumption than men because less alcohol is metabolized in their stomachs, permitting more to get into the bloodstream and from there to the liver. Proportionately, more alcoholic women die from cirrhosis than do alcoholic men. Death rates for all kinds of cirrhosis, however, are two to three times higher for men than for women.

Alcoholics with liver disease may also have kidneys enlarged with increased fat, protein, and water content. Alcoholics suffer from tissue loss in their kidneys at about 20 times the rate of nonalcoholics—7 to 14 percent of all alcoholics have kidney tissue loss.

According to NIAAA Director Enoch Gordis, "Prevention remains the key approach to the problem of ALD, but research provides hope that at least some effects may be reversible even after the disease has become established." Total abstinence from alcohol is the treatment for fatty liver, alcoholic hepatitis, and alcoholic cirrhosis. For those whose medical condition has greatly deteriorated, liver transplants are the only alternative.

Digestion and Nutrition

The gastrointestinal (GI) tract is also affected by chronic heavy drinking. Alcohol interferes with the functions of all parts of the GI tract, from the mouth to the large intestine. Alcohol abuse may damage the linings of the stomach and intestines, causing the formation of bleeding ulcers and producing abdominal pain and discomfort.

Alcohol abuse has also been linked to the development of cancers of the tongue, larynx, pharynx, and esophagus. Recent studies have also established an association between alcohol use and colorectal cancer. An estimated 75 percent of esophageal cancers in the United States are attributable to chronic, excessive alcohol consumption. Nearly 50 percent of cancers of the mouth, pharynx, and larynx are associated with heavy drinking. If a person drinks and smokes, the increased risk is even more dramatic.

Alcohol is very high in empty calories; it contains no vitamins and minerals (except for a very few in beer). A drinker who gets many calories from alcohol intake usually has little appetite for other food. Primary malnutrition may then develop, sometimes producing vomiting and diarrhea. The severe vomiting that accompanies heavy drinking causes tears in the lining of the esophagus.

Secondary malnutrition occurs when the food consumed is poorly digested. Alcohol reduces the absorption of food through the lining of the small intestine and interferes with the absorption of amino acids, glucose, zinc, and vitamins. Lack of vitamin B_1 (thiamine) can cause Wernicke's disease, damaging the brain.

Heavy alcohol consumption is a primary cause of chronic pancreatitis. More than 75 percent of patients with this disease have a history of heavy drinking, frequently for 5 to 10 years before the symptoms appear. Pancreatitis causes severe abdominal pain, often accompanied by nausea, vomiting, and fever. Medical researchers do not yet fully understand how alcohol damages the pancreas.

Cancer

Alcohol is thought to facilitate the delivery of carcinogens within the body and to impair the body's natural immune responses. Excessive alcohol consumption increases the risk of several digestive tract cancers. Cancers of the mouth, tongue, pharynx, and esophagus are found more frequently in alcoholics than in non-alcoholics. The exact cause of these cancers is unknown, but alcohol is known to irritate the mucous membranes.

Several studies have suggested that alcohol consumption increases the risk of breast cancer. In "Alcohol and Breast Cancer in Women: A Pooled Analysis of Cohort Studies" (Stephanie A. Smith-Warner et al., *Journal of the American Medical Association*, February 18, 1998), the research results show that the risk of breast cancer increases linearly with an increasing intake of alcohol. That is, the more a woman drinks, the higher her risk for breast cancer. The authors conclude that reducing alcohol consumption is a potential means of reducing breast cancer risk.

Prolonged, heavy drinking has been associated with many cases of primary liver cancer. However, it is cirrhosis of the liver that is thought to induce the cancer.

The Brain

Alcohol injures brain tissue and alters activity in the parts of the brain that control memory, emotion, and thinking. Alcoholic dementia is characterized by physical changes in the brain along with an intellectual decline, including a loss of abstract thinking and problem solving abilities, difficulty in swallowing, and difficulty in manipulating objects.

Alcohol-related dementia accounts for nearly 20 percent of admissions to state mental health facilities. Chronic brain injury caused by alcohol is second only to Alzheimer's disease as a known cause of mental deterioration in adults. Many of the symptoms of both diseases are the same—loss of abstract thought processes, deterioration of speech, and loss of coordination. Alzheimer's disease, however, is always progressive (it continually gets worse), while alcohol-related dementia may not be. If a person stops drinking, the deterioration may stop and, in some instances, reverse itself.

Korsakoff's psychosis, a brain syndrome caused by chronic alcohol dependency, is a permanent state of cognitive dysfunction—the inability to remember recent events or to learn new information. Previously learned information may interfere with new learning, and it may also be difficult to recall events that occurred before the onset of the psychosis. A related syndrome called Wernicke's disease manifests itself with vision problems, ataxia (loss of the ability to control muscle movement), and confusion. Unlike Korsakoff's psychosis, Wernicke's disease can be reversed with an adequate intake of thiamine (vitamin B_1).

The Heart

Chronic alcohol consumption may contribute to congestive heart failure through toxic effects on the heart muscle, producing cardiomyopathy (degeneration of the heart muscle). This damage to the heart muscle often follows a long period (about a decade) of heavy drinking. Symptoms of damage include shortness of breath, ankle swelling, unusual fatigue, and eventual heart failure.

Alcohol also interferes with the ability of the heart to contract, and heartbeat irregularities (cardiac arrhythmias) are common in alcoholics. Chronic alcohol consumption is also associated with a significant increase in hypertension (high blood pressure) and may be an important factor in ischemic heart disease (deficient blood circulation to the heart) and cerebrovascular disorders, including stroke.

Other Muscles

Alcohol abuse can also weaken skeletal muscles. Alcoholic myopathy (muscle disease) may be either acute or chronic. Symptoms of acute myopathy include muscular pain, tenderness, and weakness, usually confined to one limb or a group of muscles. The chronic form of the disease causes a slow progression of muscular weakness and atrophy (wasting). Mild symptoms occur in about one-third of alcoholics. Alcohol also affects involuntary smooth-muscle contractions, a primary reason why it was once used as a treatment for controlling the contractions of the uterus in premature labor.

The Blood and Immune Systems

Alcohol can cause blood abnormalities. Common problems are anemia, a decrease in hemoglobin in the blood, and abnormally enlarged red blood cells. When an alcohol-dependent person stops drinking, his or her red blood cells do not return to their normal size for a minimum of four months.

Another common blood abnormality in alcoholics is a lowered white blood cell count. White blood cells are an integral part of the body's defense against disease, and a low white cell count leaves many alcoholics more susceptible to infectious diseases than non-alcoholics. Additionally, chronic alcohol use reduces the T-cell population in lymphoid tissue and the spleen. T-cells are white blood cells that are key to the immune response.

There is no evidence of a direct association between alcohol use and AIDS (acquired immunodeficiency syndrome); however, alcohol use may increase risk-taking behavior such as sharing drug needles or having sex without taking proper precautions to reduce the risk of contracting sexually transmitted infections.

Some evidence also shows that long-term drinking can produce irregularities in blood sugar levels, although alcoholism has not been tied conclusively to diabetes.

Sexuality and Reproduction

An estimated 70 to 80 percent of alcoholics suffer from impotence and/or reduced sexual drive. Male alcoholics tend to have much lower levels of testosterone, the principal male hormone, than do non-alcoholics, while their levels of estrogen, a female hormone, are increased. As a result, in advanced stages of alcoholism, males may show certain female physical characteristics.

In premenopausal women, chronic heavy drinking can contribute to a variety of reproductive disorders. These include the cessation of menstruation, irregular menstrual cycles, failure to ovulate, early menopause, and increased risk of spontaneous abortions. These dysfunctions can be caused directly by the interference of alcohol with the hormonal regulation of the reproductive system, or indirectly through other disorders associated with alcohol abuse, such as liver disease, pancreatic disease, malnutrition, or fetal abnormalities.

Fetal Alcohol Syndrome

Fetal Alcohol Syndrome (FAS) was first described in studies in France in 1968 and in the United States in 1973. Since these early investigations, research results have shown that alcohol consumption during pregnancy can cause fetal defects. In spite of the known risks, many mothers-to-be continue to drink during their pregnancies. In 1999 the *National Household Survey on Drug Abuse* (NHSDA) found that of the sample of females aged 15 to 44 who were surveyed, nearly 48 percent used alcohol in the month prior to the survey. Of the pregnant women in this sample, 13.8 percent used alcohol during the prior month, 3.4 percent engaged in binge drinking, and 0.5 percent were heavy alcohol users. (See Table 3.6.)

Data from the 1997 NHSDA show that about three times as many unmarried pregnant women as married pregnant women used alcohol in the month prior to that survey. Additionally, a higher percentage of women used alcohol in their first trimester of pregnancy (when they may not have known they were pregnant) than in their second or third trimesters. Additionally, pregnant women with less than a high school education were more likely to have used alcohol in the past month than those who were high school graduates.

TABLE 3.6

Percentages of females (aged 15 to 44) reporting past month use of alcohol, by pregnancy status, 1999

	Total[1]	PREGNANCY STATUS	
		Pregnant	Not Pregnant
Alcohol	47.8	13.8	49.3
"Binge"alcohol use[2]	18.8	3.4	19.4
Heavy alcohol use[2]	3.8	0.5	4.0

[1] Estimates in the total column are for all females aged 15 to 44, including those with missing pregnancy status.

[2] "Binge" alcohol use is defined as drinking five or more drinks on the same occasion on at least 1 day in the past 30 days. By "occasion" is meant at the same time or within a couple hours of each other. Heavy alcohol use is defined as drinking five or more drinks on the same occasion on each of 5 or more days in the past 30 days; all heavy alcohol users are also "binge" alcohol users.

SOURCE: "Table G.27. Percentages Reporting Past Month Use of Tobacco and Alcohol Among Females Aged 15 to 44, by Pregnancy Status: 1999," in *Summary of Findings from the 1999 National Household Survey on Drug Abuse,* Substance Abuse and Mental Health Services Administration, Rockville, MD, 2000

FAS symptoms are often difficult to recognize, especially when a child is young. Estimates of the incidence of FAS range from about 0.2 to 1.0 cases per 1,000 births in the general population. Within the population of alcoholic mothers, however, the rate of FAS increases substantially, ranging from 23 to 29 FAS babies for every 1,000 births.

The U.S. Department of Health and Human Services reports a steady increase in the reported rate of FAS, from 0.1 cases per 1,000 births in 1979 to nearly 0.7 per 1,000 births in 1993. The increase does not necessarily mean that more babies are being born with FAS, but may mean that health care practitioners now have greater awareness and understanding of the syndrome. These data are the latest available. However, in April 2001, the Centers for Disease Control and Prevention opened the National Center on Birth Defects and Developmental Disabilities, which will collect data on FAS on a regular basis.

The basic criteria for recognizing FAS include:

- Prenatal and postnatal (before and after birth) growth retardation;

- Heart malformations;

- Central nervous system problems, including developmental delays, behavioral dysfunction, intellectual impairments, and poor development of the skull or brain;

- Characteristic facial features, including small eye openings, a broad thin upper lip, and a flattened nose bridge and mid-face.

CONSEQUENCES OF FAS. FAS produces a wide variety of learning and behavioral problems that appear at different stages of development. FAS infants have trouble with feeding. FAS children are often irritable and have trouble sleeping and eating; fine motor skills do not develop properly.

In the first years of school, most children with FAS are diagnosed with attention deficit hyperactivity disorder (ADHD) because of their high activity level, short attention span, and poor short-term memory. Many have special educational needs, even if their IQs are within the normal range. They are impulsive and have problems communicating and socializing with other students. In adolescence, FAS children frequently have difficulties with decision-making, abstract thinking, and problem solving. Many drop out of school.

Adults who were born with FAS often have continued problems integrating into society. They may have difficulty finding and holding jobs because of their poor impulse control, lack of social skills and, often, functional illiteracy. As a result, they have an increased risk of drug abuse and criminal activity.

IS THERE A SAFE LEVEL OF ALCOHOL USE DURING PREGNANCY? Although women who drink heavily during pregnancy are at risk of giving birth to babies with FAS, not all who consume alcohol, even in large amounts, deliver FAS babies. Scientists still do not know whether the risk of harm to the fetus is associated with the total amount of alcohol consumed by the mother during pregnancy or with the maximum level of alcohol in the blood at given times. In other words, is FAS caused by continuous drinking during pregnancy, or is it caused by getting very drunk once or twice? It is generally believed that even moderate maternal drinking may pose risks to a developing fetus, perhaps resulting in subtle effects in the offspring.

Researchers do not agree on whether there is a threshold below which it is safe to drink without endangering the fetus. Given the relatively high levels of alcohol at which significant defects have been documented, some clinicians find it difficult to justify the need for complete abstinence. In "Prenatal Alcohol Exposure and Neurobehavioral Development—Where Is the Threshold?" (*Alcohol Health & Research World*, 1994), Joseph Jacobson and Sandra Jacobson note, however, that "because alcohol exposure has no apparent benefit for the developing fetus and is not necessary for the health and well being of the mother, some clinicians and health officials have argued that the best policy is to advise pregnant women not to drink at all during pregnancy."

The surgeon general has advised that all pregnant women abstain from drinking throughout pregnancy because there is no way to determine which babies may be at risk for damage from very low levels of alcohol exposure.

INTERACTION WITH OTHER DRUGS

Because alcohol is easily available and such an accepted part of American social life, people often forget that it is a drug. When someone takes a medication while drinking alcohol, he or she is actually taking two drugs. Alcohol taken in combination with other drugs, such as an illegal drug like cocaine, an over-the-counter drug like cough medicine, or a prescription drug like an antibiotic, may counteract the effectiveness of a prescribed medication or may be potentially dangerous.

To promote the desired chemical or physical effects, a medication must be absorbed into the body and must reach its site of action. Alcohol may prevent an appropriate amount of the medication from reaching its site of action. In other cases, alcohol can alter the drug's effects once it reaches the site. Alcohol interacts negatively with more than 150 medications. Table 3.7 shows some possible effects of combining alcohol and other types of drugs.

Sometimes persons who abuse alcohol use illegal drugs as well. Abusing more than one drug at the same time is called polyabuse. M. Fischman and C. Johanson, in *Pharmacological Aspects of Drug Dependence: Towards an Integrated Neurobehavior (Approach Handbook of Experimental Pharmacology)* (Springer-Verlag, New York, 1996), report that 60 percent to 90 percent of cocaine abusers also use alcohol. Using alcohol and cocaine together creates a new metabolite (a substance produced by the chemical reactions of the body) that scientists have found increases the addictiveness of cocaine and the likelihood of cocaine-induced brain damage.

Sometimes persons drink alcoholic beverages while they are using prescription or over-the-counter (nonprescription) drugs. More than 10,000 prescription drugs and approximately 300,000 nonprescription drugs are available in the United States. H.D. Holder, in *Effects of Alcohol, Alone and in Combination with Medications* (Prevention Research Center, Walnut Creek, California, 1992), estimates that alcohol-medication interactions may be a factor in at least 25 percent of all emergency room admissions.

The Food and Drug Administration recommends that anyone who regularly has three alcoholic drinks a day should check with a doctor before taking aspirin, Tylenol, or any other over-the-counter painkiller. Combining alcohol with aspirin, ibuprofen (such as AdvilR or MotrinR), or related pain relievers may promote stomach bleeding. Combining alcohol with acetaminophen (such as Tylenol) may promote liver damage.

Elderly Americans, who consume 25 to 30 percent of all prescription medications, are particularly vulnerable. They take more drugs, are more likely than younger people to suffer medication side effects, and their bodies have greater difficulty absorbing both medicine and alcohol.

ALCOHOL-RELATED DEATHS

In 1998 the tenth-leading cause of death in the United States was cirrhosis and other chronic liver diseases.

TABLE 3.7

Interactions between alcohol and medications

Substances	Interactions
Antidepressants	Alcohol slows the breakdown of these drugs and increases their toxicity.
Acetaminophen (aspirin substitute)	Alcohol can increase this painkiller's toxic effects on the liver.
Aspirin	Aspirin may increase stomach irritation caused by alcohol.
Antihistamines	Alcohol increases the sedative effects of these drugs.
Sedatives	Alcohol increases the effects of many of these drugs and can be dangerously toxic.
Antacid histamine blockers	These drugs can interfere with the metabolism of alcohol, making it more intoxicating.

SOURCE: Prepared by staff of Information Plus

Slightly over 25,000 people died from cirrhosis. While not all of these deaths were alcohol-related, about 44 percent of cirrhosis deaths in North America are listed as such. A study in Canada, however, found that about 80 percent of all cirrhosis deaths and 90 percent of cirrhosis deaths in men ages 35 to 60 were alcohol-related. Cirrhosis is the second most common cause of death (after traffic accidents) in which alcohol is the main factor.

According to the National Center for Health Statistics' *National Vital Statistics Reports* (July 24, 2000), 19,515 persons died of alcohol-induced causes in the United States in 1998. This category includes deaths from dependent use of alcohol, nondependent use of alcohol, and accidental alcohol poisoning. It excludes accidents, homicides, and other causes indirectly related to alcohol use as well as deaths due to fetal alcohol syndrome.

Specific Causes of Death

J. Michael Gaziano and his research associates ("Light-to-Moderate Alcohol Consumption and Mortality in the Physicians' Health Study Enrollment Cohort," *Journal of the American College of Cardiology*, 2000) examined the relationship between light-to-moderate alcohol consumption in men and specific causes of death. Table 3.8 compares the total mortality (number of deaths) of groups of light-to-moderate drinkers with a reference group of nondrinkers. The total mortality of the nondrinker reference group is shown as 1.00. Any number less than one indicates a decrease in total mortality compared to the nondrinker group. (On average, persons in these groups live longer than those in the nondrinker group.) Any number higher than one indicates an increase in total mortality compared to the nondrinker group. (On average, persons in these groups do not live as long as those in the nondrinker group.) Table 3.8 shows a decrease in total mortality among those who consumed less than two drinks per day. Those who consumed two or

TABLE 3.8

Relative risk of total mortality according to level of alcohol consumption

	Number of Drinks						
	Rarely/Never	1-3/Month	1/Week	2-4/Week	5-6/Week	1/Day	≥2/Day
Age adjusted cases (n=3216)	723	338	335	477	284	889	170
RR (95% CI)[1]	1.00 (Ref)	0.88 (0.78-1.01)	0.74 (0.65-0.84)	0.73 (0.65-0.82)	0.76 (0.77-0.88)	0.85 (0.77-0.94)	1.13 (0.96-1.34)
RF[2] adjusted cases (n=2912)							
RR (95% CI)[1]	1.00 (Ref)	0.86 (0.75-0.99)	0.74 (0.65-0.85)	0.77 (0.68-0.87)	0.78 (0.67-0.90)	0.82 (0.74-0.92)	0.95 (0.79-1.14)

[1]p for nonlinear association < 0.05.
[2]Adjusted for age and other coronary risk factors, including smoking, diabetes, exercise and body mass index.
CI= confidence interval; RF= risk factor; RR= relative risk.

SOURCE: J. Michael Gaziano et al., "Table 3: RR of Total Mortality According to Level of Alcohol Consumption," from "Light-to-Moderate Alcohol Consumption and Mortality in the Physicians' Health Study Enrollment Cohort," reprinted with permission from the American College of Cardiology, *Journal of the American College of Cardiology*, 2000, vol. 35, no. 1, pp 96–105

TABLE 3.9

Relative risk of cardiovascular disease mortality according to level of alcohol consumption

	Number of Drinks						
	Rarely/Never	1-3/Month	1/Week	2-4/Week	5-6/Week	1/Day	≥2/Day
Total cardiovascular disease							
Age adjusted (n=1450)[1]	1.00 (Ref)	0.95 (0.79-1.14)	0.78 (0.65-0.95)	0.75 (0.64-0.89)	0.76 (0.62-0.93)	0.77 (0.66-0.89)	0.92 (0.71-1.20)
RF[2] adjusted (n=1328)[1]	1.00 (Ref)	0.93 (0.76-1.13)	0.78 (0.65-0.95)	0.79 (0.67-0.95)	0.81 (0.65-1.00)	0.74 (0.63-0.87)	0.76 (0.57-1.01)
Myocardial infarction							
Age adjusted (n=514)[1]	1.00 (Ref)	0.82 (0.61-1.11)	0.68 (0.50-0.91)	0.62 (0.47-0.82)	0.57 (0.40-0.80)	0.50 (0.39-0.65)	0.70 (0.44-1.10)
RF[2] adjusted (n=480)[1]	1.00 (Ref)	0.86 (0.63-1.16)	0.64 (0.47-0.89)	0.68 (0.51-0.91)	0.61 (0.42-0.88)	0.53 (0.41-0.69)	0.60 (0.36-0.98)
Stroke							
Age adjusted (n=150)	1.00 (Ref)	1.09 (0.59-1.99)	0.70 (0.36-1.36)	0.62 (0.34-1.16)	1.02 (0.54-1.93)	1.30 (0.84-2.01)	0.96 (0.40-2.31)
RF[2] adjusted (n=136)	1.00 (Ref)	0.95 (0.49-1.83)	0.62 (0.30-1.28)	0.59 (0.30-1.15)	1.10 (0.58-2.11)	1.21 (0.76-1.94)	0.84 (0.34-2.04)
Other cardiovascular disease							
Age adjusted (n=786)	1.00 (Ref)	1.03 (0.79-1.34)	0.89 (0.69-1.16)	0.89 (0.71-1.13)	0.89 (0.67-1.17)	0.91 (0.74-1.11)	1.12 (0.79-1.60)
RF[2] adjusted (n=712)	1.00 (Ref)	0.99 (0.75-1.31)	0.93 (0.71-1.22)	0.94 (0.73-1.20)	0.94 (0.70-1.26)	0.84 (0.67-1.05)	0.89 (0.61-1.30)

[1]p for linear trend < 0.05. [2]Adjusted for age and other coronary risk factors, including smoking, diabetes, exercise and body mass index.
RF= risk factor.

SOURCE: J. Michael Gaziano et al., "Table 4: RR (95% CI) of CVD Mortality According to Level of Alcohol Consumption," from "Light-to-Moderate Alcohol Consumption and Mortality in the Physicians' Health Study Enrollment Cohort," reprinted with permission from the American College of Cardiology, *Journal of the American College of Cardiology*, 2000, vol. 35, no. 1, pp. 96–105

more drinks per day had approximately the same total mortality as the nondrinker group.

Table 3.9 compares the drinker and nondrinker groups with respect to death from cardiovascular disease (CVD). The data show that alcoholic beverages lower the risk of dying from a myocardial infarction (MI, or heart attack) in those who consume from one drink per month to two or more drinks per day. The greatest protective effect for stroke is seen at one to four drinks per week.

Table 3.10 compares the drinker and nondrinker groups with respect to death from cancer. Although it appears that there is a protective effect for all cancers at one to six drinks per week, the data points really encompass the wide ranges shown in parentheses. The authors conclude that the results suggest little or no increased overall cancer risk among light-to-moderate male drinkers and do not suggest that alcohol has a protective effect against cancer.

Deborah A. Dawson ("Alcohol Consumption, Alcohol Dependence, and All-Cause Mortality," *Alcoholism: Clinical and Experimental Research*, 2000) studied the effects of alcohol consumption and alcohol dependence on the risk of mortality. The results of this study support the findings of Gaziano and colleagues (and previous studies as well), which suggest a reduced risk of mortality in light-to-moderate drinkers. However, Dawson found that those dependent on alcohol do not share the protective effect of light and moderate drinking, and very heavy drinkers have a significantly increased risk of mortality relative to those whose abstain from alcohol.

TABLE 3.10

Relative risk of cancer mortality according to level of alcohol consumption

				Number of Drinks			
	Rarely/Never	1-3/Month	1/Week	2-4/Week	5-6/Week	1/Day	≥2/Day
Total Cancer							
Age adjusted (n=944)[1]	1.00 (Ref)	0.91 (0.71-1.17)	0.77 (0.60-0.98)	0.83 (0.67-1.03)	0.83 (0.64-1.08)	1.00 (0.83-1.20)	1.53 (1.15-2.05)
RF[2] adjusted (n=864)[1]	1.00 (Ref)	0.92 (0.72-1.19)	0.71 (0.55-0.93)	0.83 (0.66-1.04)	0.81 (0.62-106)	0.92 (0.76-1.12)	1.18 (0.87-1.62)
Lung Cancer							
Age adjusted (n=207)[1]	1.00 (Ref)	0.55 (0.28-1.08)	0.68 (0.37-1.21)	1.02 (0.64-1.62)	1.03 (0.60-1.77)	1.22 (0.82-1.81)	2.53 (1.47-4.35)
RF[2] adjusted (n=181)	1.00 (Ref)	0.50 (0.25-1.01)	0.63 (0.34-1.16)	0.97 (0.60-1.57)	0.85 (0.48-1.52)	0.89 (0.58-1.36)	1.30 (0.72-2.34)
Colon Cancer							
Age adjusted (n=112)	1.00 (Ref)	1.17 (0.57-2.42)	0.82 (0.37-1.76)	1.12 (0.30-2.11)	0.79 (0.35-1.82)	1.37 (0.79-2.38)	1.26 (0.47-3.37)
RF[2] adjusted (n=101)	1.00 (Ref)	1.28 (0.61-2.69)	0.88 (0.40-1.93)	1.08 (0.55-2.11)	0.84 (0.36-1.97)	1.21 (0.66-2.20)	1.01 (0.34-3.02)
Prostate Cancer							
Age adjusted (n=78)[1]	1.00 (Ref)	0.56 (0.21-1.50)	0.10 (0.01-0.71)	0.93 (0.46-1.89)	0.70 (0.28-1.77)	1.19 (0.67-2.13)	0.80 (0.24-2.71)
RF[2] adjusted (n=74)[1]	1.00 (Ref)	0.56 (0.21-1.50)	0.09 (0.01-0.71)	0.86 (0.42-1.78)	0.69 (0.27-1.76)	1.01 (0.55-1.85)	0.69 (0.20-2.37)
Pancreatic Cancer							
Age adjusted (n=85)	1.00 (Ref)	0.95 (0.46-1.97)	0.50 (0.21-1.19)	0.44 (0.20-0.97)	0.70 (0.31-1.58)	0.73 (0.40-1.32)	1.62 (0.69-3.81)
RF[2] adjusted (n=78)	1.00 (Ref)	0.89 (0.40-1.97)	0.50 (0.20-1.25)	0.52 (0.23-1.15)	0.72 (0.30-1.72)	0.79 (0.42-1.49)	1.77 (0.73-4.29)
Hematologic Cancer[3]							
Age adjusted (n=137)	1.00 (Ref)	0.98 (0.54-1.81)	1.15 (0.66-2.00)	0.90 (0.53-1.53)	0.76 (0.39-1.49)	0.62 (0.37-1.05)	0.79 (0.31-2.04)
RF[2] adjusted (n=124)	1.00 (Ref)	1.09 (0.59-2.04)	1.08 (0.60-1.96)	0.87 (0.49-1.53)	0.70 (0.33-1.45)	0.67 (0.38-1.15)	0.70 (0.24-2.02)
Other Cancers							
Age adjusted (n=325)	1.00 (Ref)	1.08 (0.73-1.62)	0.88 (0.59-1.33)	0.72 (0.49-1.06)	0.84 (0.54-1.31)	0.97 (0.71-1.33)	1.56 (0.96-2.56)
RF[2] adjusted (n=296)	1.00 (Ref)	1.10 (0.72-1.68)	0.78 (0.50-1.22)	0.75 (0.50-1.12)	0.88 (0.56-1.39)	0.98 (0.70-1.37)	1.28 (0.74-2.21)

[1]p for nonlinear association < 0.05. [2]Adjusted for age and other coronary risk factors, including smoking, diabetes, exercise and body mass index. [3]Hematologic includes both lymphoma and leukemia.
RF= risk factor.

SOURCE: J. Michael Gaziano et al., "Table 5: RR (95% CI) of Cancer Mortality According to Level of Alcohol Consumption," from "Light-to-Moderate Alcohol Consumption and Mortality in the Physicians' Health Study Enrollment Cohort," reprinted with permission from the American College of Cardiology, *Journal of the American College of Cardiology*, 2000, vol. 35, no. 1, pp. 96–105

Michael Thun and colleagues ("Alcohol Consumption and Mortality Among Middle-Aged and Elderly U.S. Adults," *The New England Journal of Medicine*, 1997) calculated death rates according to self-reported alcohol consumption in 500,000 Americans 30 years of age or older. Figure 3.7 shows death rates for men and for women by amount of alcohol consumption. These results are similar to other research results and show that alcohol has a protective effect against all cardiovascular diseases, which extends to six or more drinks per day. The protective effect of alcohol against all causes of death is shown by a U-shaped curve, in which the protective effect is greatest at a consumption of less than one drink daily to about two drinks per day. As expected, alcohol increases death rates for alcohol-augmented conditions at about two drinks per day.

Alcohol and Trauma

The association between alcohol and accidents has long been recognized. A passage in an Egyptian papyrus from 1500 B.C. warned that drinking could lead to falls and broken bones. Short-term effects of alcohol include diminished motor coordination and balance, and impaired attention and judgment. Alcohol contributes to many types of accidents: motor vehicle, pedestrian, and work-related accidents, as well as drownings and fires.

MOTOR VEHICLE ACCIDENTS. The National Highway Traffic Safety Administration (NHTSA) of the U.S. Department of Transportation defines a traffic accident as alcohol-related if either the driver or an involved pedestrian had blood alcohol concentrations (BAC) of 0.01 percent (grams per deciliter) or greater. In many states, persons with a BAC of 0.10 percent or higher are considered intoxicated.

As of April 2001, 23 states, in addition to the District of Columbia, had a legal BAC limit of 0.08 (see Table 3.11). The NHTSA surveyed the effectiveness of the lower level in 11 states that had sufficient experience with 0.08 BAC laws. The data from all cases studied show that the rate of alcohol involvement in fatal crashes has been declining. Additionally, at least nine independent studies have shown that 0.08 BAC laws are associated with reductions in alcohol-related fatalities.

In October 2000 President Bill Clinton signed legislation that requires states to lower their BAC limit to 0.08 percent or lose a portion of their federal highway funding. Beginning in October 2003, states that have not passed such laws will lose 2 percent of their highway funding. Each year, the amount of funding lost will increase by 2 percent, until 2006, when states will lose a maximum of 8 percent of these funds annually.

FIGURE 3.7

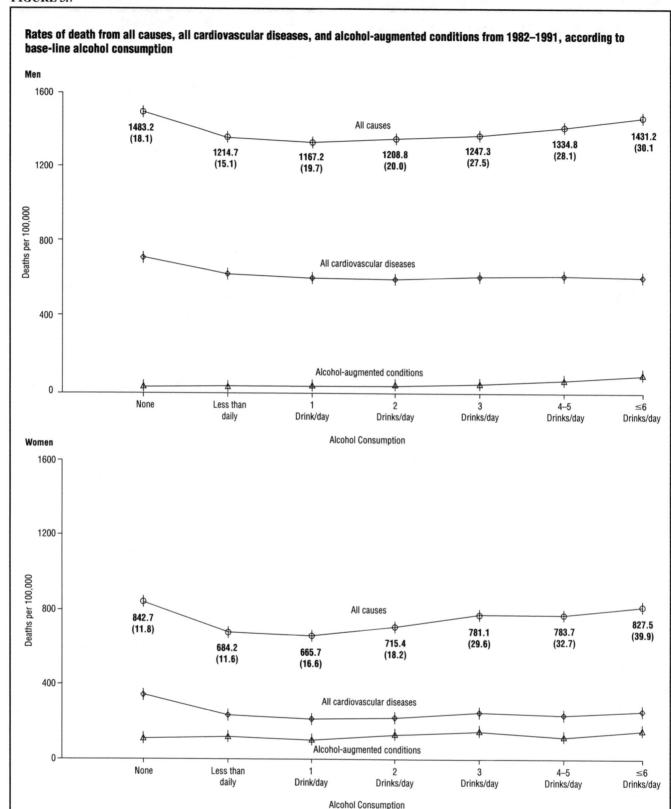

Rates of death from all causes, all cardiovascular diseases, and alcohol-augmented conditions from 1982–1991, according to base-line alcohol consumption

Alcohol-augmented conditions are cirrhosis and alcoholism, alcohol-related cancers, breast cancer in women, and injuries and other external causes. "Less than daily" alcohol consumption was defined as drinking three or more times per week but less than one drink per day. The numbers in parentheses are the standard errors of the rates of death from all causes.

SOURCE: Michael J. Thun, Richard Penn, Alan D. Lopez, Jane H. Monaco, Jane Henley, Clark W. Heath and Richard Doll "Figure 1. Rates of Death from All Causes, All Cardiovascular Diseases, and Alcohol-Augmented Conditions from 1982 to 1991, According to Base-Line Alcohol Consumption," in "Alcohol Consumption and Mortality Among Middle-Aged and Elderly U.S. Adults," in *The New England Journal of Medicine,* Vol. 337, No. 24, December 11, 1997. Copyright © 1997 Massachusetts Medical Society. All rights reserved.

As of April 2001, 41 states and the District of Columbia also had administrative license revocation (ALR) laws, which require prompt, mandatory suspension of drivers' licenses for failing or refusing to take the BAC test. This immediate suspension, prior to conviction and independent of criminal procedures, is invoked right after arrest. ALR laws are intended to protect the public from impaired drivers.

Motor vehicle crashes are the leading cause of death for those aged 5 to 44 ("Deaths: Preliminary Data for 1999," *National Vital Statistics Reports*, Centers for Disease Control and Prevention, June 2001). In 1999, 41,542 people were killed in traffic accidents, 15,786 of them in alcohol-related accidents. Alcohol-related accidents accounted for 7 percent of all crashes in 1999 and 38 percent of fatal crashes, down sharply from 51 percent in 1987 (*Traffic Safety Facts 1999—Alcohol*, National Highway Traffic Safety Administration, Washington, D.C., 2000).

As Table 3.12 shows, drivers aged 21 to 44 were most likely to be involved in fatal crashes involving a BAC of 0.10 or more in both 1989 and 1999. While the fatal crash rate of drivers with a BAC of 0.10 or greater dropped in all age groups between 1989 and 1999, the sharpest decline (30 percent) occurred among young drivers ages 16 to 20. This decline may be due to greater compliance in 1999 with laws making 21 the minimum drinking age, which went into effect in all states in 1988. (The NHTSA estimates that these laws have saved 19,121 lives between 1975 and 1999.) The rate of fatal crashes for drivers over the age of 64 years and having a BAC of 0.10 or greater declined by 29 percent. In 1999 the percentage of male drivers involved in fatal crashes with a BAC of 0.10 was twice that of females (20 percent versus 10 percent, respectively).

Of the types of drivers involved in fatal crashes in 1999 most likely to have a BAC of 0.10 or more, motorcyclists came in first (28 percent), followed by drivers of light trucks (20 percent), passenger cars (17 percent), and large trucks (1 percent). (See Table 3.12.) Approximately three out of ten pedestrians (31 percent) and 22 percent of pedalcyclists (for example, bicyclists) killed in fatal crashes in 1999 had a BAC of 0.10 or more (see Table 3.13).

FALLS, DROWNING, AND BURNS. Data from the National Center for Injury Prevention and Control (NCIPC), an office of the Centers for Disease Control and Prevention, show that excessive alcohol consumption is a significant factor in more than 100,000 deaths in the United States each year. Falls were the second leading cause of unintentional injury death in the United States in 1998. In 1998 more than 12,595 fatalities resulted from unintentional falls (*National Vital Statistics Reports*, July 2000). According to the NCIPC *Fact Book for the Year 2000*, alcohol consumption is involved in 18 to 53 percent of nonfatal falls each year.

TABLE 3.11

States with 0.08 Blood Alcohol Concentration (BAC) Laws

Alabama	Maryland
Arkansas	Nebraska
California	New Hampshire
District of Columbia	New Mexico
Florida	North Carolina
Georgia[1]	Oregon
Hawaii	Rhode Island
Idaho	Texas
Illinois	Utah
Kansas	Vermont
Kentucky	Virginia
Maine	Washington

[1] Under Georgia law, a prosecutor only has to prove a person was driving with a BAC of 0.10 or greater to get a conviction. At BACs of 0.08 or greater but less than 0.10, there is a presumption the defendant was driving under the influence of alcohol, but that presumption can be rebutted by the defendant.

SOURCE: "States with 0.08 BAC Laws," in *Blood Alcohol Concentration Limits for Enforcement of Impaired Driving Laws–U.S. States–2001*, Indiana Prevention Resource Center at Indiana University, 2001.

Drowning was the fifth leading cause of unintentional injury death in the United States in 1998. The NCIPC reports that alcohol is involved in 40 to 50 percent of drowning deaths among teenage boys. In addition, alcohol use is involved in about 50 percent of all deaths associated with water recreation, such as boating. An NCIPC analysis of survey data on alcohol use during recreational boating revealed that 31 percent of respondents who operated motorboats reported doing so at least once while alcohol-impaired. Most of these respondents were male and between the ages of 25 and 34.

Forty-nine states and the District of Columbia impose a BAC limit of 0.10 or lower for boating and drinking. All states have boating-under-the-influence laws (National Association of State Boating Law Administrators, *Reference Guide to State Boating Laws*, fifth ed., 1999). Penalties for boating while intoxicated include fines, imprisonment, substance abuse or boating safety classes, suspension of boat operating privileges, and, in some states, the loss of one's motor vehicle driver's license.

Fires were the fifth leading cause of unintentional injury death in the United States in 1998, killing 3,363 people (*National Vital Statistics Reports*, July 2000). According to a study cited by the NCIPC in their *Fact Book for the Year 2000*, alcohol contributes to 40 percent of deaths in residential fires.

ALCOHOL, VIOLENCE, AND CRIME

Experts agree that heavy use and/or abuse of alcohol often go hand-in-hand with violence and crime. Alcohol reduces inhibitions and distorts judgment, and drunkenness is sometimes accepted by society as an excuse or partial excuse for out-of-control behavior.

TABLE 3.12

Alcohol involvement for drivers in fatal crashes, 1989 and 1999

Drivers Involved in Fatal Crashes	1989 Number of Drivers	1989 Percentage with BAC 0.10 g/dl or Greater	1999 Number of Drivers	1999 Percentage with BAC 0.10 g/dl or Greater	Change in Percentage, 1989-1999
Total Drivers					
Total*	60,435	24	56,352	17	-29%
Drivers by Age Group (Years)					
16–20	9,442	20	7,973	14	-30%
21–24	7,723	35	5,620	27	-23%
25–34	15,928	32	11,734	24	-25%
35–44	10,106	25	11,023	21	-16%
45–64	10,240	17	12,292	13	-24%
Over 64	5,431	7	6,559	5	-29%
Drivers by Sex					
Male	45,448	27	40,900	20	-26%
Female	14,054	14	14,792	10	-29%
Drivers by Vehicle Type					
Passenger Cars	35,204	24	27,806	17	-29%
Light Trucks	15,579	28	19,801	20	-29%
Large Trucks	4,903	3	4,847	1	-67%
Motorcycles	3,182	40	2,515	28	-30%

*Numbers shown for groups of drivers do not add to the total number of drivers due to unknown or other data not included.

SOURCE: "Table 3: Alcohol Involvement for Drivers in Fatal Crashes, 1989 and 1999" in *Traffic Safety Facts 1999—Alcohol,* National Highway Traffic Safety Administration, National Center for Statistics and Analysis, Washington, D.C., 2000

TABLE 3.13

Alcohol involvement for nonoccupants killed in fatal crashes, 1989 and 1999

Nonoccupant Fatalities	1989 Number of Nonoccupant Fatalities	1989 Percentage with BAC 0.10 g/dl or Greater	1999 Number of Nonoccupant Fatalities	1999 Percentage with BAC 0.10 g/dl or Greater	Change in Percentage, 1989-1999
Pedestrian Fatalities by Age Group (Years)					
16–20	383	37	273	34	-8%
21–24	387	50	231	44	-12%
25–34	1,165	54	620	46	-15%
35–44	911	49	905	50	+2%
45–64	1,244	38	1,175	37	-3%
Over 64	1,467	9	1,084	10	+11%
Total*	**6,556**	**32**	**4,906**	**31**	**-3%**
Pedalcyclist Fatalities					
Total	832	14	750	22	+57%

*Includes pedestrians under 16 years old and pedestrians of unknown age.

SOURCE: "Table 5: Alcohol Involvement for Nonoccupants Killed in Fatal Crashes, 1989 and 1999" in *Traffic Safety Facts 1999—Alcohol,* National Highway Traffic Safety Administration, National Center for Statistics and Analysis, Washington, D.C., 2000

Alcohol and Crime

In 1997, 20.4 percent of federal prisoners and 37.2 percent of state prisoners reported that they had been under the influence of alcohol alone (not in combination with any other drug) when they committed their offenses. In general, the more violent the crime, the more likely it was that the offender was using alcohol. (See Table 3.14.)

Although studies have shown that large proportions of serious criminal offenders drink heavily before com-

mitting crimes, implying a correlation between the two activities, there is still no definitive evidence that drinking alcohol actually causes crimes. Offenders may first decide to commit their crimes and then get drunk to control their fears, or they may impulsively commit crimes after they have been drinking.

In addition to major crimes, other alcohol-related offenses take up much of the time and budgets of law enforcement officials. In *Crime in the United States, 1999*

TABLE 3.14

Alcohol or drug use at time of offense of state and federal prisoners, by type of offense, 1997

| | Estimated number of prisoners[a] | | Percent of prisoners who reported being under the influence at time of offense | | | | | |
| | | | Alcohol | | Drugs | | Alcohol or drugs | |
Type of offense	State	Federal	State	Federal	State	Federal	State	Federal
Total	1,046,705	88,018	37.2%	20.4%	32.6%	22.4%	52.5%	34.0%
Violent offenses	494,349	13,021	41.7%	24.5%	29.0%	24.5%	51.9%	39.8%
Murder	122,435	1,288	44.6	38.7	26.8	29.4	52.4	52.4
Negligent manslaughter	16,592	53	52.0	...	17.4	...	56.0	...
Sexual assault[b]	89,328	713	40.0	32.3	21.5	7.9	45.2	32.3
Robbery	148,001	8,770	37.4	18.0	39.9	27.8	55.6	37.6
Assault	97,897	1,151	45.1	46.0	24.2	13.8	51.8	50.5
Other violent	20,096	1,046	39.6	32.2	29.0	15.9	48.2	37.2
Property offenses	230,177	5,964	34.5%	15.6%	36.6%	10.8%	53.2%	22.6%
Burglary	111,884	294	37.2	...	38.4	...	55.7	...
Larceny/theft	43,936	414	33.7	...	38.4	...	54.2	...
Motor vehicle theft	19,279	216	32.2	...	39.0	...	51.2	...
Fraud	28,102	4,283	25.2	10.4	30.5	6.5	42.8	14.5
Other property	26,976	757	36.0	22.8	30.6	16.4	53.2	34.6
Drug offenses	216,254	55,069	27.4%	19.8%	41.9%	25.0%	52.4%	34.6%
Possession	92,373	10,094	29.6	21.3	42.6	25.1	53.9	36.0
Trafficking	117,926	40,053	25.5	19.4	41.0	25.9	50.9	35.0
Other drug	5,955	4,922	29.9	19.7	47.1	17.1	59.2	29.0
Public-order offenses	103,344	13,026	43.2%	20.6%	23.1%	15.6%	56.2%	30.2%
Weapons	25,642	6,025	28.3	23.0	22.4	24.4	41.8	37.1
Other public-order	77,702	7,001	48.1	18.5	23.3	8.1	60.9	24.1

...Too few cases in the sample to permit calculation.
[a] Based on cases with valid offense data.
[b] Includes rape and other sexual assault.

SOURCE: "Alcohol or drug use at time of offense of state and federal prisoners, by type of offense, 1997" in *Substance Abuse and Treatment, State and Federal Prisoners, 1997,* Bureau of Justice Statistics, Washington, D.C., 1999

(Washington, D.C., 2000), the Federal Bureau of Investigation (FBI) reported that in 1999, there were

- 1,549,500 arrests for driving under the influence (alcohol and other drugs).
- 683,600 arrests for liquor law violations (open container laws, selling to minors, etc.).
- 673,400 arrests for drunkenness.
- 30,800 arrests for vagrancy, which is often an alcohol-related offense.

Alcohol and Violence Among College Students

During the college years, alcohol abuse becomes a serious problem for some students. Such abuse often begins or accelerates during the college years. In fact, the most commonly abused drug among college students is alcohol.

A study conducted by C.A. Presley and colleagues ("Weapon Carrying and Substance Abuse among College Students," *Journal of the American College of Health*, 1997) revealed that male college students who carried weapons consumed significantly more alcohol than male college students who did not carry weapons. The results for armed versus unarmed female students did not show this difference in alcohol intake.

In 1997, according to the *College Alcohol Survey, 1979–1997* (George Mason University, Fairfax, Virginia;

and West Chester University, West Chester, Pennsylvania), 64 percent of all violent-behavior incidents on campus involved alcohol. Over half (58 percent) of campus property damage was the result of alcohol.

Alcohol and Domestic Violence

Researchers have also reported a high level of alcohol use in cases of domestic violence, both child abuse and partner abuse. In addition, heavy drinking, both before and after marriage, is considered a risk factor for potential domestic violence. In some cases of spousal (husband or wife) abuse, both partners have been drinking before the violence occurs; in other cases, only one spouse has been drinking.

According to L. Greenfeld, in *Alcohol and Crime: An Analysis of National Data on the Prevalence of Alcohol Involvement in Crime* (Bureau of Justice Statistics, 1998), slightly over two-thirds of victims of intimate violence (assaults by current or former spouses, boyfriends, or girlfriends) reported that alcohol was a factor in the attack. For assaults by spouses, about three-fourths reported their husbands or wives had been drinking. (See Table 3.15.)

Studies of domestic violence have found that alcohol does not cause abuse but that it is used by abusers as an excuse for the behavior. Alcohol abusers often claim that they are out of character when they are drunk and, consequently, not accountable for their behavior. Dr. Richard

TABLE 3.15

Violent crime and alcohol or drug abuse

	Percent of violent victimizations with offender using alcohol
All victims	37%
Intimate*	67
Nonmarital relatives	50
Acquaintances	38
Strangers	31

	Percent of spouse violence victimizations involving substance use
Alcohol only	65%
Drugs only	5
Both alcohol and drugs	11
Either alcohol or drugs	<1
Neither alcohol nor drugs	19

*Includes current or former spouse, boyfriend, and girlfriend.

SOURCE: *Alcohol and Crime,* Bureau of Justice Statistics, Washington, D.C., 1998

Gelles, the Joanne T. and Raymond B. Welsh Chair of Child Welfare and Family Violence at the University of Pennsylvania, observes in *Intimate Violence in Families,* third edition (Sage Publications, Thousand Oaks, California, 1997):

[In many cases,] individuals who wish to carry out a violent act become intoxicated in order to carry out the act. Alcohol leads to violence because it sets off a primary conflict over drinking that can extend to arguments over spending money, working, and sex. In these cases, drinking may serve as a trigger for long-standing marital disputes and disagreements, the existence of suitable and acceptable justifications for violence serves to normalize and neutralize the violence. These justifications also may play a causal role in family violence by providing, in advance, an excuse for behavior that is normally prohibited by societal and familial norms and standards.

CHAPTER 4

ALCOHOL ABUSE AND ADDICTION

Alcohol is an addictive substance, but not everyone who drinks alcohol becomes addicted. About 90 percent of those who drink alcohol do not become alcoholics. Scientists cannot explain what individual traits account for the difference, but they suspect that a wide variety of factors may make a person more susceptible to addictions of all kinds.

Alcoholism was recognized as a disease over two hundred years ago. In 1785 Benjamin Rush, a signer of the Declaration of Independence and the first physician-general of Washington's Continental Army, wrote an essay on "the effects of ardent spirits," calling intemperance a disease and an addiction. Throughout the 19th century, physicians considered intemperance a disease. Opposition to the disease concept was widespread, however—especially among those who advocated a moralistic view of alcoholism. The Temperance Movement, for example, believed that alcoholism could be cured through personal dedication or as part of a commitment to God.

THE DEFINITION OF ALCOHOLISM

The definition of alcoholism has been revised and refined as scientists have learned more about it. Most people consider an alcoholic to be someone who drinks too much and cannot control his or her drinking. Alcoholism, however, does not merely refer to heavy drinking or getting drunk a certain number of times. The diagnosis of alcoholism applies only to those who show specific symptoms of addiction.

Drs. Robert Morse and Daniel Flavin, writing for the Joint Committee of the National Council on Alcoholism and Drug Dependence and the American Society of Addiction Medicine ("The Definition of Alcoholism," *Journal of the American Medical Association,* 1992), define alcoholism as:

[A] primary, chronic disease with genetic, psychosocial, and environmental factors influencing its development

and manifestations. The disease is often progressive and fatal. It is characterized by impaired control over drinking, preoccupation with the drug alcohol, use of alcohol despite adverse consequences, and distortions in thinking, most notably denial. Each of these symptoms may be continuous or periodic.

"Primary" refers to alcoholism as a disease independent from any other psychological disease (e.g., schizophrenia), rather than a symptom of some other underlying disease. "Adverse consequences" for an alcoholic can include physical illness (liver disease, withdrawal symptoms, etc.), psychological problems, interpersonal difficulties (e.g., marital problems or domestic violence), and problems at work.

This definition of alcoholism incorporated "denial" as a major concept for the first time. Denial includes a number of psychological maneuvers by the drinker to avoid the fact that alcohol is the cause of his or her problems. Family and friends may reinforce an alcoholic's denial by covering up his or her drinking (e.g., calling an employer to say the alcoholic has the flu rather than a hangover). Denial is almost always a major obstacle to recovery.

The Institute of Medicine (IOM) has defined addiction as a brain disease "manifested by a complex set of behaviors that are the result of genetic, biological, psychological, and environmental interactions."

ALCOHOL ABUSE OR ALCOHOLISM?

The American Psychiatric Association, publisher of the *Diagnostic and Statistical Manual of Mental Disorders* (DSM), which has been revised several times, first defined alcoholism in 1952 (DSM-I). DSM-III introduced the subdivisions of alcohol abuse and alcohol dependence (alcoholism). The former involves a compulsive use of alcohol and impaired social or occupational functioning, while the latter includes evidence of physical tolerance or withdrawal symptoms when the drug is stopped. The

TABLE 4.1

Diagnostic criteria for alcohol dependence*

Symptoms	Diagnostic and Statistical Manual of Mental Disorders, 3rd Ed., Revised (DSM-III-R)	Diagnostic and Statistical Manual of Mental Disorders, 4th Ed., (DSM-IV)	International Classification of Diseases (ICD), 10th Ed.
	A. At least three of the following:	A. A maladaptive pattern of alcohol use, leading to clinically significant impairment or distress as manifested by three or more of the following occurring at any time in the same 12-month period:	A. Three or more of the following have been experienced or exhibited at some time during the previous year.
Tolerance	(1) Marked tolerance—need for markedly increased amounts of alcohol (i.e., at least 50 percent increase) in order to achieve intoxication or desired effect, or markedly diminished effect with continued use of the same amount of alcohol	(1) Need for markedly increased amount of alcohol to achieve intoxication or desired effect; or markedly diminished effect with continued use of the same amount of alcohol	(1) Evidence of tolerance, such that increased doses are required in order to achieve effects originally produced by lower doses
Withdrawal	(2) Characteristic withdrawal symptoms for alcohol (3) Alcohol often taken to relieve or avoid withdrawal symptoms	(2) The characteristic withdrawal syndrome for alcohol; or alcohol (or a closely related substance) is taken to relieve or avoid withdrawal symptoms	(2) A physiological withdrawal state when drinking has ceased or been reduced as evidenced by: the characteristic alcohol withdrawal syndrome, or use of alcohol (or a closely related substance) to relieve or avoid withdrawal symptoms
Impaired Control	(4) Persistent desire or one or more unsuccessful efforts to cut down or control drinking (5) Drinking in larger amounts or over a longer period than the person intended	(3) Persistent desire or one or more unsuccessful efforts to cut down or control drinking (4) Drinking in larger amounts or over a longer period than the person intended	(3) Difficulties in controlling drinking in terms of onset, termination, or levels of use
Neglect of Activities	(6) Important social, occupational, or recreational activities given up or reduced because of drinking	(5) Important social, occupational, or recreational activities given up or reduced because of drinking	(4) Progressive neglect of alternative pleasures or interests in favor of drinking; or
Time Spent Drinking	(7) A great deal of time spent in activities necessary to obtain alcohol, to drink, or to recover from its effects	(6) A great deal of time spent in activities necessary to obtain alcohol, to drink, or to recover from its effects	A great deal of time spent in activities necessary to obtain alcohol, to drink, or to recover from its effects
Inability to Fulfill Roles	(8) Frequent intoxication or withdrawal symptoms when expected to fulfill major role obligations at work, school, or home; or	None	None
Hazardous Use	When drinking is physically hazardous	None	None
Drinking Despite Problems	(9) Continued drinking despite knowledge of having a persistent or recurring social, psychological, or physical problem that is caused or exacerbated by alcohol use	(7) Continued drinking despite knowledge of having a persistent or recurring physical or psychological problem that is likely to be caused or exacerbated by alcohol use	(5) Continued drinking despite clear evidence of overtly harmful physical or psychological consequences
Compulsive Use	None	None	(6) A strong desire or sense of compulsion to drink
Duration Criterion	B. Some symptoms of the disturbance have persisted for at least one month or have occurred repeatedly over a longer period of time	B. No duration criterion separately specified. However, three or more dependence criteria must be met within the same year and must occur repeatedly as specified by duration qualifiers associated with criteria (e.g., "often," "persistent," "continued")	B. No duration criterion separately specified. However, three or more dependence criteria must be met during the previous year.
Criterion For Subtyping Dependence	None	With physiological dependence: Evidence of tolerance or withdrawal (i.e., any of items A(1) or A(2) above are present) Without physiological dependence: No evidence of tolerance or withdrawal (i.e., any of items A(1) or A(2) above are present)	None

NOTE: Information pertaining to this table remains the same in the DSM-IV-TR (Text Revision) as in the DSM-IV.

*The DSM-III-R, DSM-IV, and ICD-10 refer to substance dependence. These criteria have been adapted to focus solely on alcohol.

SOURCE: "Diagnostic Criteria for Alcohol Abuse and Dependence," in *Alcohol Alert*, no. 30 PH 359, October 1995

TABLE 4.2

Diagnostic criteria for alcohol abuse/harmful use of alcohol*

DSM-III-R Alcohol Abuse

    A. A maladaptive pattern of alcohol use indicated by at least one of the following:

        (1) continued use despite knowledge of having a persistent or recurrent social, occupational, psychological, or physical problem that is caused or exacerbated by use of alcohol

        (2) drinking in situations in which use is physically hazardous

    B. Some symptoms of the disturbance have persisted for at least one month, or have occurred repeatedly over a longer period of time.

    C. Never met the criteria for alcohol dependence.

DSM-IV Alcohol Abuse

    A. A maladaptive pattern of alcohol use leading to clinically significant impairment or distress, as manifested by one (or more) of the following occurring within a 12-month period:

        (1) recurrent drinking resulting in a failure to fulfill major role obligations at work, school, or home

        (2) recurrent drinking in situations in which it is physically hazardous

        (3) recurrent alcohol-related legal problems

        (4) continued alcohol use despite having persistent or recurrent social or interpersonal problems caused or exacerbated by the effects of alcohol

    B. The symptoms have never met the criteria for alcohol dependence.

ICD-10 Harmful Use of Alcohol

    A. A pattern of alcohol use that is causing damage to health. The damage may be physical or mental. The diagnosis requires that actual damage should have been caused to the mental or physical health of the user.

    B. No concurrent diagnosis of the alcohol dependence syndrome.

NOTE: Information pertaining to this table remains the same in the DSM-IV-TR (Text Revision) as in the DSM-IV.

*The DSM-III-R, DSM-IV, and ICD-10 refer to substance abuse and harmful use. These criteria have been adapted to focus solely on alcohol.

SOURCE: "Diagnostic Criteria for Alcohol Abuse and Dependence," in *Alcohol Alert*, no. 30 PH 359, October 1995

Manual's Fourth Edition, Text Revision 2000 (DSM-IV-TR), refined these definitions further.

The World Health Organization publishes the *International Classification of Diseases* (ICD), which is designed to standardize health data collection throughout the world. The Tenth Edition (ICD-10) generally defines abuse and tolerance similarly to the DSM-IV-TR. Table 4.1 compares the criteria for diagnosing alcohol dependence described in DSM-III-R (revised), DSM-IV-TR, and ICD-10, and Table 4.2 shows each report's description of alcohol abuse.

Symptoms of Alcoholism

The differences between an alcohol abuser and an alcoholic are subtle. Based on the DSM-IV-TR definition of alcohol dependence (alcoholism), a drinker would have to report, over a 12-month period, three or more of the symptoms shown in Table 4.1.

The DSM-IV-TR also describes two subtypes of dependence: with evidence of physiological dependence (withdrawal or tolerance), and without evidence of physiological dependence. In other words, a drinker can be an alcoholic without displaying physical symptoms of tolerance or withdrawal—if at least three of the other symptoms are present during a 12-month period.

- Tolerance means that increasingly larger amounts of alcohol must be consumed to achieve a desired response ("get high").

- Withdrawal consists of a set of physical symptoms that occur, usually within six to eight hours, when alcohol intake is reduced or stopped completely.

Mild symptoms of withdrawal include nausea, irritability, "the shakes," vomiting, sweating, and insomnia. In severe cases, delirium tremens (DTs) can occur within 48–96 hours after drinking stops. The individual experiences hallucinations, severe confusion, tremors, and possibly seizures, with increases in blood pressure, heart rate, and body temperature.

Without proper medical care, delirium tremens leads to death in one out of five cases. Even with medical attention, death may occur. While DTs can be medically managed, little is known about the physiological causes of this reaction.

Symptoms of Alcohol Abuse

While alcohol abuse is often considered less serious than alcoholism, Table 4.2 makes it clear that the problem is not trivial. The chronic and abusive use of alcohol can be very destructive to the life of an abuser and his or her family. An alcohol abuser

- Drinks in spite of the problems caused by his or her drinking.

- Fails to fulfill obligations at work, school, or home.

- Frequently drinks until intoxicated.

TABLE 4.3

Percentages reporting past month alcohol use, past month "binge" alcohol use, and past month heavy alcohol use, by age, 1999

Age Group	Type of alcohol use		
	any alcohol use	"Binge" alcohol use	Heavy alcohol use
12 to 17	18.6	10.9	2.5
18 to 25	58.0	38.3	13.3
26 or older	49.4	18.4	4.8

NOTE: "Binge" alcohol use is defined as drinking five or more drinks on the same occasion on at least 1 day in the past 30 days. By "occasion" is meant at the same time or within a couple of hours of each other. Heavy alcohol use is defined as drinking five or more drinks on the same occasion on each of 5 or more days in the past 30 days; all heavy alcohol users are also "binge" alcohol users.

SOURCE: "Table G.34. Percentages Reporting Past Month Alcohol Use, Past Month "Binge" Alcohol Use, and Past Month Heavy Alcohol Use Among Persons Aged 12 to 17, by Geographic Characteristics: 1999," "Table G.35. Percentages Reporting Past Month Alcohol Use, Past Month "Binge" Alcohol Use, and Past Month Heavy Alcohol Use Among Persons Aged 18 to 25, by Geographic Characteristics: 1999," and "Table G.36. Percentages Reporting Past Month Alcohol Use, Past Month "Binge" Alcohol Use, and Past Month Heavy Alcohol Use Among Persons Aged 26 or Older, by Geographic Characteristics: 1999," in *Summary of Findings from the 1999 National Household Survey on Drug Abuse*, Substance Abuse and Mental Health Services Administration, Rockville, MD, 2000

• May have legal problems (drinking and driving arrests, etc.) or injuries as a result of being intoxicated.

Other warning signs of alcohol abuse include the need to drink before facing certain situations, frequent drinking sprees, a steady increase in intake, solitary drinking, early-morning drinking, and the occurrence of blackouts. Blackouts for heavy drinkers are not episodes of passing out, but are periods the drinkers cannot remember later, even though they appeared to be functioning at the time. Blackouts may be an early sign of alcoholism.

PREVALENCE OF ALCOHOL ABUSE AND ALCOHOLISM

The Strategic Plan 2001–2005 of the National Institute on Alcohol Abuse and Alcoholism (NIAAA) notes that nearly 14 million Americans—1 in every 13 adults—have alcohol-abuse or alcohol-dependence problems. Laura Jean Bierut and her colleagues, in "Co-occurring Risk Factors for Alcohol Dependence and Habitual Smoking" (*Alcohol Research & Health,* 2000), note that 19 percent of American men and 8 percent of American women have been diagnosed with alcohol dependence. As Table 4.3 shows, persons ages 18–25 are far more likely to engage in alcohol use, binge alcohol use, or heavy alcohol use than persons 12–17, or those 26 and older.

Statistically, the current level of alcohol abuse and dependence among older adults is low—between 2 and 4 percent. Although overall drinking decreases as people age, some elderly persons become dependent on alcohol.

As their metabolism slows, they become more sensitive to alcohol and get drunk more quickly. In addition, the average senior citizen takes five different medications—most of which do not mix with alcohol.

Retirement, loss of a spouse, and loneliness can spark drinking. Figure 4.1 shows the stress levels of elderly nonproblem and problem drinkers and the social resources available to help them. As shown, problem drinkers have far fewer resources available.

CAUSES OF ALCOHOL ABUSE AND ALCOHOLISM

Scientists, physicians, and social workers agree that there is no one cause of alcoholism. It is the result of a complex mix of sociological, psychological, and behavioral factors that interact with a person's genetic makeup, resulting in a complex mix.

Sociological Factors

Why do some national, religious, and cultural groups have high rates of alcoholism while others do not? The NIAAA suggests that low alcoholism rates occur in certain groups because the drinking customs and sanctions (permissions) are well established and consistent with the rest of the culture. Conversely, multicultural populations with mixed feelings about alcohol and no common rules tend to have high alcoholism rates.

Additionally, certain populations may be at higher or lower risk because of the way their bodies metabolize (chemically process) alcohol. For example, many Asians have an inherited deficiency of aldehyde dehydrogenase, a chemical that breaks down ethyl alcohol in the body. Without it, toxic substances build up after drinking alcohol and rapidly lead to flushing, dizziness, and nausea. Therefore, many Asians experience warning signals very early on, and are less likely to continue drinking. Conversely, research suggests that American Indians may inherit a lack of these warning signals. Therefore, they are less sensitive to the intoxicating effects of alcohol and are more likely to develop alcoholism.

Results of studies show that Irish Americans and American Indians have high rates of alcoholism, while Jewish Americans and Asian Americans have low rates of alcoholism. Overall, there is no difference in alcoholism prevalence between African Americans, whites, and Hispanics.

Psychological Factors

Psychologists and psychiatrists have described alcoholics as neurotic, unable to relate to others effectively, sexually and emotionally immature, unable to withstand frustration or tension, poorly integrated, and marked by feelings of low self-esteem. No reliable study exists to confirm these observations, which also describe many people who are not alcoholics.

FIGURE 4.1

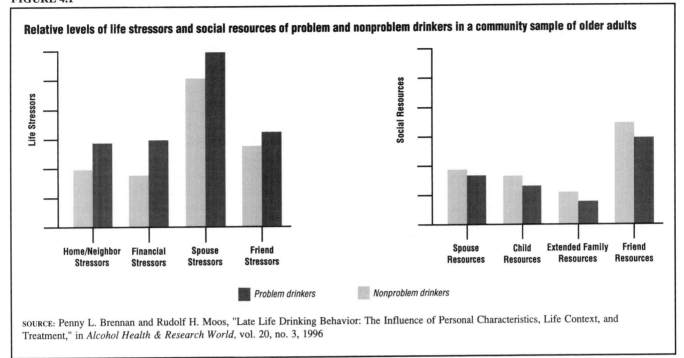

Relative levels of life stressors and social resources of problem and nonproblem drinkers in a community sample of older adults

SOURCE: Penny L. Brennan and Rudolf H. Moos, "Late Life Drinking Behavior: The Influence of Personal Characteristics, Life Context, and Treatment," in *Alcohol Health & Research World,* vol. 20, no. 3, 1996

The American Medical Association (AMA) observes that many authorities believe alcoholics suffer personality disturbances that can be traced back to childhood. The drinker may say, or even believe, that he or she is using alcohol to relieve anxiety and stress, but actually the alcohol may be covering deeper feelings, such as insecurity, rage, depression, and guilt, that originated many years earlier. Each time drinking brings relief, it becomes the solution to other stress situations. The pattern continues and eventually becomes an illness. Many authorities also believe a desire for self-destruction plays a crucial role in alcoholism.

Behavioral Factors

Alan Marlatt, a specialist in addictive behaviors at the University of Washington in Seattle, Washington, believes anyone can become an alcoholic, not just those with genetic links to alcoholism. He further believes that alcoholism is not, in itself, a disease, although it is an addiction that can lead to diseases, such as cirrhosis of the liver.

University of Pittsburgh School of Medicine psychologist Ralph Tarter has identified personality traits that he believes make a person more likely to exhibit addictive behavior such as alcoholism:

- A high activity level (not to be confused with hyperactivity in children).

- A tendency to become easily excited and hard to calm down.

- Physical signs of distress, such as sweating and upset stomach.

- An enjoyment of the taste of alcohol more acute than that of nonalcoholics.

Tarter emphasizes that these are characteristics of many types of addictive behavior, not just alcoholism. Although these traits can be inherited, environment plays a large part in determining the particular type of addiction: alcohol, nicotine, caffeine, or other drugs.

Genetic Factors

Over the past two decades, research results have shown that genetic factors contribute to alcohol dependence. Most scientists believe there is no single gene that causes alcoholism. Alcoholism is thought to be linked to various genes, each contributing alone, or in combination with others, to the risk of this disease.

Scientists at the NIAAA have launched a comprehensive multiyear study, involving hundreds of alcohol-dependent persons and thousands of their family members. The Collaborative Study on the Genetics of Alcoholism (COGA) was initiated in 1989, with the goal of detecting and mapping genes that confer susceptibility to alcohol dependence and related disorders, including dependence on other drugs such as nicotine.

The principal investigator for this study, Henri Begleiter, believes that alcoholism is "a set of biological factors which are heavily influenced by environmental events and can lead to very different adverse reactions." These outcomes can include other addictive behaviors, such as gambling and eating disorders.

COGA researchers have concluded that the causes of alcoholism may differ significantly among individuals.

FIGURE 4.2

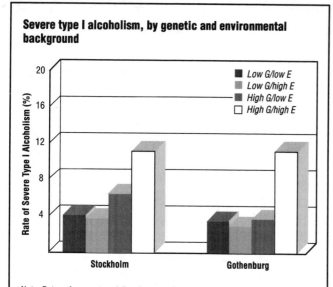

Severe type I alcoholism, by genetic and environmental background

Note: Rates of severe type I (i.e., late-onset) alcoholism among male participants in two adoption studies performed in Stockholm and Gothenburg, Sweden. Both the genetic (G) and the environmental (E) backgrounds of the adoptees were classified as either high or low risk based on multiple biological and environmental factors (e.g., the presence of type I alcoholism in the biological parents or the socio-economic status of the adoptive family). Both studies found significantly elevated rates of severe type I alcoholism only among male adoptees who both had a high genetic risk and were reared in a high-risk environment.

SOURCE: Matt McGue, "Figure 2," in "A Behavioral-Genetic Perspective on Children of Alcoholics," *Alcohol Health & Research World*, vol. 21, no. 3, 1997

By 1995 results from the study suggested that small effects from many genes combine to form the condition called alcoholism ("The Collaborative Study on the Genetics of Alcoholism," *Alcohol Health & Research World*, 1995). Researchers postulate that certain families may have a unique mix of genes that make family members more susceptible to alcoholism than others.

Results from the COGA study also showed that both alcohol dependence and habitual smoking were transmitted within families and likely had some genetic factors ("Co-occurring Risk Factors for Alcohol Dependence and Habitual Smoking," *Alcohol Research & Health*, 2000). Several regions of DNA (hereditary material) have been identified that may contain genes that confer a susceptibility to alcoholism. Some of those genes also may contribute to the risk of habitual smoking.

ADOPTION STUDIES. Adoption studies track alcoholism in adopted children whose biological mothers, fathers, or both were alcoholics. In the 1970s D. W. Goodwin and his colleagues used official records in Copenhagen, Denmark, to determine if environment made a difference in whether children born of alcoholic parents became alcohol-dependent adults. They identified three groups: (1) alcoholic parents who had given up their children for early adoption; (2) alcoholic parents who had given up one child for early adoption but had reared

another child; and (3) nonalcoholic parents who had given up children for early adoption. Group 3 was the "control" group, the standard for comparing the effects of genetics versus environment.

The researchers interviewed the adult sons and daughters of people in groups 1 and 2 to find out if rates of alcoholism in these groups were higher than in the control group. Results for daughters were inconclusive. The results for sons, however, showed a significantly greater risk for alcoholism among those adopted-away (a risk factor of 3.6) and nonadopted sons (3.4) of alcoholic fathers, compared with an assigned risk of 1.0 for the control adoptees. Put simply, sons of alcoholic parents had more than three times the risk of becoming alcoholics than sons of nonalcoholic parents, whether or not they were reared by those parents. Clearly, a genetic link existed.

Another group of researchers conducted a similar study in Stockholm, Sweden, in the 1980s. The Stockholm Adoption Study is the largest published adoption study of alcoholism. In this study the rate of alcohol abuse among male adoptees was 14.7 percent if neither biological parent abused alcohol; 22.4 percent if only the biological father abused alcohol; 26.0 percent if only the biological mother abused alcohol; and 33.3 percent if both biological parents abused alcohol. The risk for alcohol abuse among male adoptees was clearly greater among those who had one or two biological parents who were abusers.

The Stockholm researchers also analyzed data concerning female adoptees. Results showed that the rate of alcohol abuse was significantly greater among female adoptees with alcohol-abusing mothers (9.8 percent) than among female adoptees with non-alcohol-abusing mothers (2.8 percent). Paternal alcohol abuse did not affect the female adoptees' risk of alcohol abuse, suggesting that genetic effects might be gender-specific.

Results of the Stockholm Adoption Study, therefore, indicated that both male and female adoptees had a greater risk of alcohol abuse and dependence if their biological, but not adoptive, parents were alcoholic. Additionally, the researchers identified two types of alcoholism—Type I and Type II—and determined that a combination of both genetic and environmental risk factors is required for the development of the more common of the two types, Type I. The less common type, Type II, appeared to be highly heritable in men but not in women.

Adoption studies in the United States have been hampered in the past because of the privacy regulations applied to adoptions. Several studies, however, were conducted in Iowa in the late 1980s and early 1990s, using adoption records from several placement agencies. In general, these studies supported the findings of the earlier studies: there is a significantly higher risk of alcoholism for the children of alcoholics.

TABLE 4.4

Prevalence and pair resemblance for lifetime alcohol abuse and dependence among 1,514 male twin pairs

Diagnosis	Prevalence (%)[1]		Probandwise Concordance[2]		Odds Ratio[3]		Monozygotic-Dizygotic Equality of Odds Ratios[3]	Pair Correlation[4]	
	Monozygotic Twin Pairs	Dizygotic Twin Pairs	Monozygotic Twin Pairs	Dizygotic Twin Pairs	Monozygotic Twin Pairs	Dizygotic Twin Pairs		Monozygotic Twin pairs	Dizygotic Twin Pairs
DSM-III-R criteria									
Alcohol dependence	26.5	28.1	0.49	0.38	8.5	3.9	9.7	0.49	0.24
Alcohol abuse[5]	28.3	31.9[6]	0.54	0.44	10.4	4.5	12.3	0.54	0.30
Alcohol abuse or dependence	33.3	36.6	0.57	0.50	9.7	5.0	8.4	0.53	0.34
DSM-IV criteria									
Alcohol dependence	22.5	24.4	0.48	0.32	10.3	3.3	18.2	0.53	0.18
Alcohol abuse[5]	30.0	33.6[7]	0.57	0.46	11.6	4.5	15.8	0.58	0.31
Alcohol abuse or dependence	34.3	37.5	0.58	0.50	10.0	4.6	11.7	0.55	0.31

[1]Based on all male twins from complete pairs with known zygosity.
[2]Sample sizes are numbers of complete pairs and vary with diagnostic criteria. For DSM-III-R: alcohol dependence, monozygotic=444, dizygotic=363; alcohol abuse, monozygotic=477, dizygotic=415; alcohol abuse/dependence, monozygotic=560, dizygotic=475; for DSM-IV: alcohol dependence, monozygotic=378, dizygotic=316; alcohol abuse, monozygotic=505, dizygotic=436; alcohol abuse/dependence, monozygotic=576, dizygotic=475.
[3]Breslow-Day chi-square test (24), all df=1, p<0.01.
[4]Tetrachoric correlations; monozygotic pairs=861, dizygotic pairs=653.
[5]With or without dependence.
[6]Significantly greater than for monozygotic, x^2=4.3, df=1, p<0.05.
[7]Significantly greater than for monozygotic, x^2=4.6, df=1, p<0.05.

SOURCE: Carol A. Prescott and Kenneth S. Kendler, "Prevalence and Pair Resemblance for Lifetime Alcohol Abuse and Dependence Among 1,514 Male Twin Pairs," in "Genetic and Environmental Contributions to Alcohol Abuse and Dependence in a Population-Based Sample of Male Twins," *The American Journal of Psychiatry*, vol. 156, no. 1, copyright 1999, the American Psychiatric Association.

In "New Findings in the Genetics of Alcoholism" (*The Journal of the American Medical Association,* 1999) Marc A. Schuckit reviewed several adoption and twin studies on genetic links to alcoholism. Schuckit concluded that twin and adoption studies consistently confirm an important genetic influence on alcohol abuse and dependence. Schuckit also noted that alcoholism is three to four times more prevalent in first-degree relatives of alcoholics (such as brothers, sisters, or children) than in those who do not have alcoholic first-degree relatives. In identical twins born of alcoholics, this increased prevalence rate doubles.

Matt McGue, in "A Behavioral-Genetic Perspective on Children of Alcoholics" (*Alcohol Health & Research World,* 1997), discusses shared environmental factors as well as genetic factors in the development of alcohol abuse and alcohol dependence. Figure 4.2 shows data from the Stockholm Adoption Study and a replication study conducted in Gothenburg, Sweden. The data reflect the rate of severe Type I alcoholism as a function of genetic and environmental factors. The bar graphs show that there is a higher prevalence of Type I alcoholism in persons who have both high-risk genetic backgrounds and high-risk environmental factors for alcoholism.

TWIN STUDIES. Twin studies compare the risk of alcoholism in identical and fraternal twins. Identical twins are those who develop from a single egg; therefore, their genetic makeup is identical. Fraternal twins develop from two eggs that were fertilized independently of each other.

The first twin study was conducted in Sweden in the 1950s. Researchers first identified male twins, at least one of whom had a problem with alcohol. When they studied the identical twins in this group, they determined that in 61 percent of the twin pairs, both twins were dependent on alcohol. However, in only 39 percent of the fraternal twins were both twins dependent on alcohol. Thus, alcoholism appeared to have a strong genetic component. Follow-up twin studies have confirmed these findings.

Carol A. Prescott and Kenneth S. Kendler, in "Genetic and Environmental Contributions to Alcohol Abuse and Dependence in a Population-Based Sample of Male Twins" (*The American Journal of Psychiatry,* 1999), reported that genetic factors played a major role in the development of alcoholism among males in their study. From 1993 to 1996 the researchers conducted interviews with white male twins born between 1940 and 1974. The 3,473 subjects who reported any lifetime alcohol use were divided into three groups, based on the DSM-IV criteria: alcohol dependence, alcohol abuse, and nonabstainers.

Alcohol abuse and dependence among twin pairs was substantially higher among the 861 identical twins than among the 653 fraternal twins. Table 4.4 reports the results in several ways, but the information is summarized in the "odds ratio" column. The odds ratio is the risk of being affected when a twin sibling has been diagnosed with alcohol abuse or dependence, compared with the risk of being affected when a twin sibling has not been diagnosed with alcohol abuse or dependence. For instance,

TABLE 4.5

Distinguishing differences between type I and type II alcoholism*

Characteristic	Type I Alcoholism	Type II Alcoholism
Contributing factors	Genetic and environmental	Primarily genetic
Gender distribution	Affects both men and women	Affects men more often than women
Usual age of onset	After age 25	Before age 25
Common alcohol-related problems	Loss of control over drinking; binge drinking; guilt about drinking; progressive severity of alcohol abuse	Inability to abstain from alcohol; drinking frequently associated with fighting and arrests; severity of alcohol abuse usually not progressive
Characteristic personality traits	High harm avoidance and low novelty seeking; person drinks to relieve anxiety	High novelty seeking; person drinks to induce euphoria

*The characteristics listed in this table define the type I and type II prototypes that only represent the two extremes of a continuous spectrum of manifestations of alcohol abuse.

SOURCE: C. Robert Cloninger et al., "Distinguishing Differences Between Type I and Type II Alcoholism," in "Type I and Type II Alcoholism: An Update," *Alcohol Health & Research World*, vol. 20, no. 1, 1996

using DSM-IV criteria, the risk for alcohol abuse is 11.6 times higher for an identical twin whose co-twin has been diagnosed with alcohol abuse, and 4.5 times higher for a fraternal twin whose co-twin has been diagnosed with alcohol abuse, than for a twin whose co-twin is unaffected. ("Probandwise concordance," as indicated in Table 4.4, is the proportion of abuse or dependence in both twins when one is determined to be alcohol dependent or abusing. "Monozygotic" refers to identical twins; "dizygotic" refers to fraternal twins.)

Early studies of female twins did not show a strong genetic link between twin siblings for alcohol abuse or dependence. A more recent study of 1,030 female twins, however, showed a consistently higher incidence of alcoholism among both identical twins than was shown among both fraternal twins (Kenneth Kendler et al., "A Population-Based Twin Study of Alcoholism in Women," *Journal of the American Medical Association,* 1992).

Of the pairs of identical twins in which one twin displayed alcoholism with dependence-tolerance (addiction and the need to drink increasing amounts of alcohol), 26.2 percent of the co-twins were also alcoholics. Among fraternal twins, the percentage was 11.9. Among the pairs of identical twins in which one twin showed symptoms of alcoholism without dependence-tolerance, 31.6 percent of the co-twins were found to be alcohol abusers. Among fraternal twins, the percentage was 24.4. Of the identical twin pairs in which one twin had "a significant drinking problem," 46.9 percent of the co-twins also had an alcohol problem. The percentage among fraternal twins was 31.5.

Although men are far more likely to become alcoholics than are women, alcoholism is a growing problem for women. In "Genetics and Epidemiology of Alcoholism" (*American Psychiatric Press Review of Psychiatry,* 1989) Robert Cloninger et al. speculate that the reason the genetic link has begun to show up in studies of females is that the twins are younger than those used in earlier studies. "It's becoming more socially acceptable for young

women to drink as heavily as only their fathers would have in a previous generation. With this social change we are going to be seeing more women alcoholics."

GENETIC SUBTYPES OF ALCOHOLISM

Based on adoptee and twin studies, researchers have described several subtypes of alcoholism. Type I alcoholism affects both males and females, usually develops later in life, and is thought to be both genetic and environmental in cause. Type II occurs more often in men, usually develops during adolescence or young adulthood, and is primarily genetic in cause. (See Table 4.5.)

Another classification is called the Type A-Type B subtype. Type A and Type B alcoholics are defined by a range of factors, including family history of alcoholism, psychological disorders, and the severity of their alcoholism. There also appears to be a gender factor: more women than men tend to be Type A alcoholics, while men outnumber women in the Type B subtype. (See Table 4.6.)

Type A alcoholics typically have less severe dependence symptoms, and are more responsive to treatment than are Type B alcoholics. In contrast, Type B alcoholics tend to develop alcoholism at earlier ages, display more problem behaviors early in life, and have more severe dependence symptoms and alcohol-related problems (health, social, and psychological) than Type A alcoholics.

THE COST OF ALCOHOL ABUSE

Living with someone who has an alcohol problem affects every member of the family. The National Association for Children of Alcoholics (NACoA) reported in 1998 that 76 million Americans—nearly half (43 percent) of the U.S. adult population—have experienced alcoholism in the family. In "Children of Addicted Parents: Important Facts" (*NACoA Fact Sheet,* available online at http://www.nacoa.net/pdfs/addicted.pdf) the association estimates that there are more than 28 million children of alcoholics (COAs) in the United States, including nearly

TABLE 4.6

Profiles of type A and type B male and female alcoholics

Defining Characteristics of Alcoholic Subtypes	Type A	Type B
Risk Factors for Developing Alcoholism		
Familial alcoholism[1,2]	M < F	M < F
Childhood conduct disorder (e.g., behavioral problems)	M = F	M = F
Measures of personality (McAndrew Scale and MMPT[3]) [1]	M > F	M = F
Age of onset of problem drinking[1]	M < F	M = F
Alcohol and Other Substance Use		
Alcohol use (number of ounces per day[1])	M > F	M = F
Drinking to relieve negative moods and/or boredom[1]	M < F	M = F
Severity of alcohol dependence symptoms	M = F	M = F
Tranquilizer use[1]	M < F	M = F
Polydrug use	M = F	M = F
Chronicity and Consequences of Drinking		
Physical conditions resulting from alcohol use (e.g., liver disease)[1]	M < F	M = F
Physical consequences of drinking (e.g., hangovers or tremors)[1]	M < F	M = F
Social consequences of drinking (e.g., job loss or marital problems)[1]	M > F	M = F
Lifetime alcohol problems (e.g., arrests) (MAST)[2]	M = F	M > F
Number of years of heavy drinking[1,2]	M > F	M > F
Psychiatric Symptoms		
Depression symptoms (e.g., sadness)[1]	M < F	M = F
Antisocial personality (e.g., stealing or fighting)[1,2]	M > F	M > F
Anxiety symptoms (e.g., nervousness)[1]	M < F	M = F

[1]Statistically significant gender differences for type A.
[2]Statistically significant gender differences for type B.
[3]MMPT = Minnesota Multiphasic Personality Test.
[4]MAST = Michigan Alcohol Screening Test.

Note: The <, >, and = signs show how men and women compared with each other with respect to each characteristic. The findings presented are the results of a reanalysis of data presented in Babor et al. 1992.

SOURCE: Frances K. Del Boca and Michie N. Hesselbrock, "Profiles of Type A and Type B Male and Female Alcoholics," in "Gender and Alcoholic Subtypes," *Alcohol Health & Research World*, vol. 20, no. 1, 1996

11 million under the age of 18. Additionally, the article states that children of alcoholics have a risk for alcoholism and other drug abuse two to nine times greater than that of children of nonalcoholics.

Children of alcoholics are more likely to have attention-deficit hyperactivity disorder (ADHD), conduct disorders, and anxiety disorders. They tend to score lower on tests that measure cognitive and verbal skills. COAs are also more likely to be truant, repeat grades, drop out of school, or be referred to a school counselor or psychologist. Because COAs have more physical and mental-health problems, the rate of total health-care costs for COAs is 32 percent greater than that of children from nonalcoholic families.

Alcohol abuse and addiction impose a burden not only on alcoholics and their families, but also on society as a whole. Alcohol-related problems are costly in terms of medical care, treatment, rehabilitation, reduced or lost productivity, and the expenses of law enforcement.

In the *10th Special Report to the U.S. Congress on Alcohol and Health* (U.S. Department of Health and Human Services, Public Health Service, National Institutes of Health, National Institute on Alcohol Abuse and Alcoholism, June 2000) the estimated cost of alcohol abuse in 1998 was projected to have been $184.6 billion, up about 25 percent from $148 billion in 1992. In contrast, the cost of alcohol abuse in 1985 was $70.3 billion.

In 1998 health-care costs were projected to have been $26.3 billion (14.2 percent of the total alcohol-related costs). The value of lost productivity due to illness ($87.6 billion), lost future earnings due to premature death ($36.4 billion), and lost productivity due to alcohol-related crime ($10 billion) together totaled an estimated $134.2 billion—nearly 73 percent of the total alcohol-related costs. Other related alcohol-abuse costs were projected at $24 billion. (See Table 4.7.)

TREATING ALCOHOL DEPENDENCE

Alcoholism cannot be "cured," if cured means the ability to return to normal social drinking. Many authorities prefer to use the term "recovering." Once sobriety is restored, staying sober by coping with the personal and social situations that contribute to drinking is an ongoing effort.

The *10th Special Report to the U.S. Congress on Alcohol and Health* notes that more than 700,000 people in the United States receive treatment for alcoholism daily. In *Treating Alcoholism: The Illness, the Symptoms, the Treatments* (Washington, D.C., not dated) the NIAAA lists three stages of treatment:

• Detoxification, or managing acute intoxication and withdrawal to overcome the effects of drunkenness, safely rid the body of alcohol, and help the body adjust to the absence of alcohol.

TABLE 4.7

Estimated economic costs of alcohol abuse, 1992 and 1998[1]

Economic Cost	1992 ($ millions)	1998 (Projected) ($ millions)
Health care expenditures		
Alcohol use disorders: treatment, prevention, and support	5,573	7,466
Medical consequences of alcohol consumption	13,247	18,872
Total	**18,820**	**26,338**
Productivity impacts		
Lost productivity due to alcohol-related illness	69,209	87,622
Lost future earnings due to premature deaths[2]	31,327	36,499
Lost productivity due to alcohol-related crime	6,461	10,085
Total	**106,997**	**134,206**
Other impacts on society		
Motor vehicle crashes	13,619	15,744
Crime	6,312	6,328
Fire destruction	1,590	1,537
Social welfare administration	683	484
Total	**22,204**	**24,093**
Total costs	**148,021**	**184,636**

[1]The authors estimated the economic costs of alcohol abuse for 1992 and projected those estimates forward to 1998, adjusting for inflation, population growth, and other factors.
[2]Present discounted value of future earnings calculated using a 6-percent discount rate.

SOURCE: "Table 1: Estimated economic costs of alcohol abuse in the United States, 1992 and 1998," in *10th Special Report to the U.S. Congress on Alcohol and Health*, U.S. Department of Health and Human Services, Washington, DC, June 2000

- Correcting health problems that may have been brought on or aggravated by heavy drinking.

- Altering long-term behavior so that drinking patterns are not reestablished.

Some physicians prescribe the drug disulfiram (Antabuse) for daily use. If combined with alcohol, this drug produces violent headaches, nausea, and other discomforts. Many doctors, however, question the effectiveness of Antabuse, believing it to be more of a psychological than a physical agent. In other words, patients taking Antabuse who believe they will become sick if they drink alcohol tend to become ill. This drug has been marketed since 1948.

Approved by the U.S. Food and Drug Administration (FDA) in 1994, naltrexone (ReVia) has been shown to be very effective with low- and medium-risk alcohol-dependent patients when used in primary-care-based alcohol intervention programs. Naltrexone lowers the "high" associated with drinking and diminishes the craving.

Acamprosate, granted investigational drug status by the FDA and currently in late-stage clinical trials in the United States, is now being used effectively in Europe. In *Pharmacotherapy for Alcohol Dependence* (The Agency for Health Care Policy and Research, Rockville, Maryland, 1999) researchers reported that both naltrexone and acamprosate can be effective in the treatment of alcoholism. They found that the drugs can help reduce crav-

ings, decrease the frequency with which a person drinks, minimize relapse, and, in some cases, improve abstinence rates. The combination of two or more medications given simultaneously may be even more efficient.

If approved in the United States, acamprosate would be sold as Campral. The main side effect of this drug is mild diarrhea, which usually goes away after a few days. By contrast, Antabuse can be toxic if the patient drinks enough alcohol, while naltrexone can cause liver damage if prescribed in too high a dose.

Anxiety and depression are frequently associated with alcoholism. Consequently, drugs such as Prozac have also proved helpful in treating the underlying problems of those dependent on alcohol.

A Long-Term Process

In 1996 Dr. George E. Vaillant, of Harvard Medical School and Brigham and Women's Hospital in Boston, announced the results of a long-term study of recovering alcoholics. The study followed the lives and drinking patterns of problem drinkers for 50 years. Researchers found that relapse was common after two years of sobriety but was rare after five years. While 56 percent of the abusers in the study achieved two years of sobriety, 41 percent of them relapsed. Generally, an alcoholic needs to live free of symptoms for five years before he or she can be considered recovered, although alcoholism can return even after five years.

Treatment Settings

Many types of long-term treatments are available for alcohol dependence, including both inpatient and outpatient treatment programs. These programs can involve psychological approaches, medications, or a combination of the two. The *10th Special Report to the U.S. Congress on Alcohol and Health* notes that a broad range of therapies are currently available to treat alcohol dependence, including social-skills training, motivational enhancement, behavior contracting, cognitive therapy, marital and family therapy, aversion therapy, and relaxation training. Complete abstinence from alcohol and other drugs is the main goal of these treatments.

Jane Ellen Smith, in "The Community Reinforcement Approach to the Treatment of Substance Use Disorders" (*The American Journal on Addictions,* 2001), describes a program that has been repeatedly proved to be successful. The Community Reinforcement Approach (CRA) is a cognitive-behavioral treatment for all substance-use disorders, and is founded on the belief that an individual's environment can play a powerful role in encouraging or discouraging drinking and drug use. When used with alcoholics, the goal is to rearrange multiple aspects of an individual's "community" so that a sober lifestyle appears more rewarding to the alcoholic than a lifestyle including alcohol dependence.

A variation of CRA has also been developed, called Community Reinforcement and Family Training (CRAFT). This program works with family members and significant others to motivate individuals who refuse to seek treatment to do so.

Current research in treatment for alcoholism has also led to an important advance called the "brief intervention." This approach is used with patients who are at-risk or problem drinkers, but who may not be alcohol dependent. With this approach, the health-care provider identifies patients who are problem drinkers, provides them with feedback and advice on their drinking, and works toward doctor-patient agreement on an appropriate course of action to stem the problem.

In past decades treatment for alcohol abuse and dependence occurred most often within hospitals and treatment facilities (inpatient treatment). In recent years, inpatient treatment has changed dramatically. The length of stay has dropped sharply, as a result of more outpatient treatment programs and pressures from the health insurance industry to cut costs. Clients are also more likely to be addicted to other drugs along with alcohol, so treatment has shifted focus from alcohol-only dependence to dependence on alcohol and other drugs.

Depending on the situation, outpatient treatment can be intensive—up to eight hours a day, seven days a week—or periodic, with the patient coming for therapy once a week.

Alcoholics Anonymous

In 1935 two alcoholics started a group called Alcoholics Anonymous, which effectively laid the foundation for the modern self-help movement (including Alcoholics Anonymous, Al-Anon, Alateen, Overeaters Anonymous, Gamblers Anonymous, etc.). Alcoholics Anonymous (AA) groups are self-governed and independent of formal alcoholism-treatment facilities. Meetings are conducted by recovering alcoholics, without regard to formal counseling training and experience. Alcoholics Anonymous has chapters in 133 countries and an estimated worldwide membership of two million. Participation in AA or in treatment programs based on the Twelve Steps of AA is the dominant approach to alcoholism treatment in the United States today (*10th Special Report to the U.S. Congress on Alcohol and Health*).

Critical elements of the AA program include fellowship meetings, with members expected to attend 90 meetings in 90 days during the early recovery period; a sponsor system in which newly recovering alcoholics are linked to an established member for assistance and advice; and the Twelve-Step philosophy, which spells out a series of activities, or steps, that alcoholics should undertake in their recovery process.

Al-Anon and Alateen are similar programs for the families of alcoholics. In Al-Anon meetings, families learn how to deal with alcoholic family members and their own feelings about these persons. Al-Anon members also work to break their own codependent behaviors—the cycle of denial, anger, and unconscious facilitation of their family members' alcoholism. Alateen groups offer support for the children of alcoholics. Families Anonymous is generally designed to offer support for the parents of alcohol- or drug-dependent children.

Project MATCH—Patient-Treatment Matching

Caregiving professionals recognize that no single treatment is successful for all persons suffering from alcohol abuse and dependence. For many years, based on the outcomes of more than 30 studies, professionals have suggested that treatments for alcoholism should be matched to the particular characteristics of the patients. The characteristics to be considered include psychiatric and sociopathic problems, severity of alcohol involvement, cognitive impairment, and level of social support.

In 1989 the NIAAA began a study called Matching Alcohol Treatments to Client Heterogeneity (Project MATCH) to determine if the outcome of treatment is affected by matching patients to certain treatments. The study recruited 1,726 patients, of whom 75 percent were men and 25 percent women. Fifteen percent were minorities. The patients were divided into two groups: those who were recruited directly from the community on an outpatient basis, and those who had just completed an inpatient or intensive day hospital treatment (the "aftercare" group). Each patient was randomly assigned to one of three treatments (all of which were conducted by qualified therapists):

- Twelve-Step Facilitation (TSF)—Twelve weekly sessions that explained Twelve-Step principles and introduced the first five Steps. Patients were encouraged to join Alcoholics Anonymous and become involved in its activities, in addition to the TSF program.

- Cognitive-Behavioral Therapy (CBT)—Twelve weekly sessions in which therapists taught skills that could help patients cope with situations and moods that are known to trigger relapses.

- Motivational Enhancement Therapy (MET)—Four sessions over a span of twelve weeks in which therapists used motivational psychology techniques. Patients were encouraged to consider their situations and how alcohol had affected their lives, develop a plan to stop drinking, and implement the plan.

Patient characteristics were studied to evaluate whether treatments that were appropriately matched to a patient's needs produced better outcomes than treatments that were not matched. By the late 1990s, the data were analyzed and the results showed that patient-treatment matching had little effect on the outcomes. In all the programs, patients decreased their drinking days per month to 6, compared with 25 before treatment. While almost all patients reported both heavy drinking and recurrent problems when they entered the project, only 50 percent reported these problems one year after treatment.

Only four patient characteristics (out of a potential 21 characteristics) showed any differences in outcome when matched with particular programs:

- Alcohol dependence—In the aftercare group, individuals highly dependent on alcohol benefited more from the Twelve-Step treatment than from cognitive-behavioral treatment. The reverse was true for patients with low alcohol dependence.

- Psychopathology—In the outpatient group, individuals without mental and behavioral disorders benefited more from the Twelve-Step treatment than from cognitive-behavioral treatment.

- Anger—In the outpatient group, individuals with high levels of anger benefited more from the motivational enhancement treatment than from the other two treatments.

- Social Network Support for Abstinence—Individuals without strong social support networks benefited more from the Twelve-Step treatment than from motivational enhancement therapy.

CHAPTER 5

TOBACCO—WHAT IT IS AND WHAT IT DOES

The use of tobacco in North America dates back to pre-Columbian days. After Christopher Columbus landed in the New World in 1492, he and later European settlers were introduced to tobacco by the American Indians. The use of tobacco products, especially cigarettes, became increasingly widespread in the United States well into the 20th century. Smoking was often associated with romance, relaxation, and adventure; movie stars oozed glamour on screen while smoking, and movie tough guys were never more masculine than when lighting up. Songs like "Smoke Gets in Your Eyes" topped the hit parade. Smoking became a rite of passage for many young males, and as women began to attain increased opportunities, they, too, began smoking.

During the past few decades, however, opposition to tobacco use has grown. Health authorities warn of the dangers of smoking or chewing tobacco, and nonsmokers object to "secondhand smoke"—because of both the smell and the health dangers of breathing smoke from other people's cigarettes. Today, a smoker is more likely to ask for permission before lighting up, and the answer is often "no."

PHYSICAL PROPERTIES OF NICOTINE

Tobacco is a plant native to the Western Hemisphere. It contains nicotine, a drug classified as a stimulant, although it has some depressive effects as well. Nicotine is a poisonous alkaloid that is the major psychoactive (mood-altering) ingredient in tobacco. (Alkaloids are carbon- and nitrogen-containing compounds that are found in some families of plants. They have both poisonous and medicinal properties.)

Chemically, nicotine is a very complex substance, and its effects on the body are also complex. It affects the brain and central nervous system, as well as the hypothalamus and pituitary glands of the endocrine (hormone) system. Nicotine easily crosses the blood-brain barrier and accumulates in the brain—faster than caffeine and heroin, but slower than Valium (a medication used to treat anxiety).

In the brain, nicotine imitates the actions of the hormone epinephrine (adrenaline) and the neurotransmitter acetylcholine, both of which heighten awareness. Nicotine also triggers the release of dopamine, which enhances feelings of pleasure, and endorphins, "the brain's natural opiates," which have a calming effect.

As mentioned previously, nicotine acts as both a stimulant and a depressant. By stimulating certain nerve cells in the spinal cord, nicotine relaxes the nerves and slows some reactions, such as knee-jerk reflex. Small amounts of nicotine stimulate some nerve cells, but these cells are depressed by large amounts. In addition, nicotine stimulates the brain cortex (the outer layer of the brain) and affects the functions of the heart and lungs.

THE ADDICTIVE NATURE OF NICOTINE

In *The Health Consequences of Smoking—Nicotine Addiction: A Report of the Surgeon General* (U.S. Department of Health and Human Services, Rockville, Maryland, 1988) researchers examined the addictiveness of tobacco. They determined that the pharmacologic (chemical/physical) effects and behavioral processes that contribute to tobacco addiction are very similar to those that contribute to addiction to drugs such as heroin and cocaine. Nicotine is considered as addictive as cocaine and heroin. Compared with some other drugs, nicotine seems to create dependence rather quickly in many users.

Cigarette smoking results in rapid distribution of nicotine throughout the body, reaching the brain within 10 seconds of inhalation. However, the intense effects of nicotine disappear in a few minutes, causing smokers to continue smoking frequently throughout the day in order to maintain its pleasurable effects and to prevent withdrawal. Tolerance develops after repeated exposure to nicotine, and higher doses are required to produce the

FIGURE 5.1

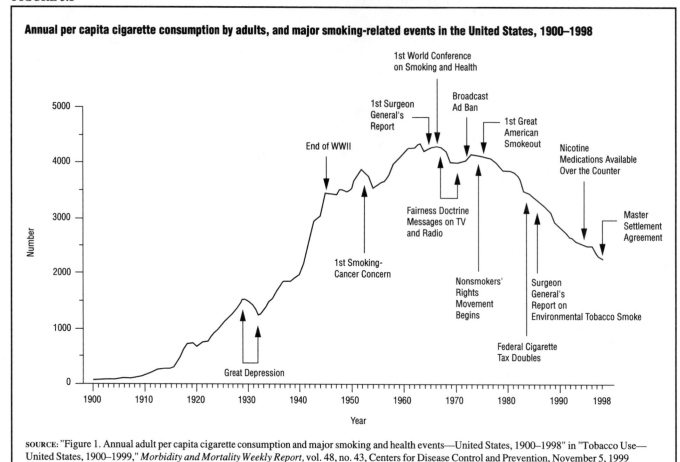

Annual per capita cigarette consumption by adults, and major smoking-related events in the United States, 1900–1998

SOURCE: "Figure 1. Annual adult per capita cigarette consumption and major smoking and health events—United States, 1900–1998" in "Tobacco Use—United States, 1900–1999," *Morbidity and Mortality Weekly Report,* vol. 48, no. 43, Centers for Disease Control and Prevention, November 5, 1999

same initial stimulation. Because nicotine is metabolized fairly quickly, disappearing from the body in a few hours, some tolerance is lost overnight. Smokers often report that the first cigarette of the day is the best. The more cigarettes smoked during the day, the more tolerance develops, and the less effect subsequent cigarettes have.

Stopping nicotine use causes a withdrawal syndrome that may begin within a few hours after the last cigarette and last a month or even more. Symptoms include craving, irritability, attentional deficits, interruption of thought processes, sleep disturbances, and increased appetite. High levels of craving for tobacco may continue for six months or longer.

Most smokers believe that they are addicted to tobacco. An October 1999 Gallup Poll asked smokers if they considered themselves addicted to cigarettes. A large majority (72 percent) felt that they were addicted. In another survey, respondents rated tobacco as more quickly addictive and harder to quit than cocaine or heroin. However, in the 1999 Gallup Poll survey, 77 percent of smokers responded that they believed they could quit if they made a decision to do so.

As with other drugs, however, not all users of tobacco become dependent. When David Mendez, Assistant Pro-

fessor of Public Health at the University of Michigan, was analyzing smoking statistics from surveys conducted for the Centers for Disease Control and Prevention, he discovered that 18 percent of the country's estimated 45 million smokers said they smoked, but not on a daily basis. Some researchers even believe that occasional smokers constitute a growing trend.

Is There a Genetic Basis for Addiction?

Recently, scientists identified a gene that appears to influence whether some people are more likely to become addicted to nicotine than others. Rachel F. Tyndale of the University of Toronto, Ontario, and her colleagues compared 244 habitual smokers with 184 people who had tried tobacco but had not become addicted. They found that those in the nonaddicted group were much more likely to have inactive versions of this gene. Among those who smoked regularly, those with inactive versions of the gene smoked fewer cigarettes.

Previous studies have identified one or two other genes thought to play a role in nicotine addiction. Some genetics experts believe that the basis of nicotine addiction is more complex than simply one or two genes. Others believe that genetic factors account for only about half of the susceptibility to nicotine addiction; the other half depends on the environment in which someone is raised.

Results from the Collaborative Study on the Genetics of Alcoholism (COGA) support the hypothesis that some common genetic factors are involved in the susceptibility for developing both alcohol and nicotine addiction ("Co-occurring Risk Factors for Alcohol Dependence and Habitual Smoking," *Alcohol Research & Health*, 2000). Moreover, twin studies have supported the role of common genetic factors in the development of both disorders.

Nicotine May Not Be the Only Addictive Substance in Cigarettes

Research results suggest that nicotine may not be the only psychoactive (mind-altering) ingredient in tobacco. Some as-yet-unknown compound in cigarette smoke decreases the levels of monoamineoxidase (MAO), an enzyme responsible for breaking down the brain chemical dopamine. The decrease in MAO results in higher dopamine levels and may be another reason that smokers continue to smoke—to sustain the high dopamine levels that result in pleasurable effects and the desire for repeated cigarette use.

TRENDS IN TOBACCO USE

Cigarettes

CENTERS FOR DISEASE CONTROL AND PREVENTION. According to the Centers for Disease Control and Prevention (CDC; *Morbidity and Mortality Weekly Report,* 1999) the consumption of cigarettes, the most widely used tobacco product, has decreased over the past generation. After increasing more or less consistently for 60 years, the per capita consumption of cigarettes peaked in the 1960s and early 1970s at about 4,000 cigarettes per year. Since 1974 the per capita consumption has consistently declined each year, reaching a low of an estimated 2,261 cigarettes per capita in 1998. (See Figure 5.1.)

Figure 5.2 shows that in 1997 the percentage of male smokers (27.6) was higher than the percentage of women smokers (22.1). The percentages of both male and female smokers have dropped somewhat consistently since 1964, when a surgeon general's report concluded that cigarette smoking is a cause of lung and laryngeal cancer in men, a probable cause of lung cancer in women, and the most important cause of chronic bronchitis in both genders.

THE NATIONAL HOUSEHOLD SURVEY ON DRUG ABUSE. Each year the Substance Abuse and Mental Health Services Administration (SAMHSA) of the U.S. Department of Health and Human Services (HHS) surveys American households on drug use by means of *The National Household Survey on Drug Abuse* (NHSDA). In 1985, 78 percent of Americans (150 million) reported smoking cigarettes at some time during their lives, and 39 percent (75 million) were current smokers (had smoked within the month prior

FIGURE 5.2

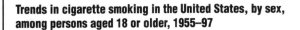

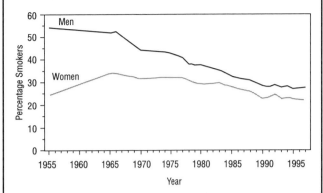

Trends in cigarette smoking in the United States, by sex, among persons aged 18 or older, 1955–97

Note: Before 1992, current smokers were defined as persons who reported having smoked ≥100 cigarettes and who currently smoked. Since 1992, current smokers were defined as persons who reported having smoked ≥100 cigarettes during their lifetime and who reported now smoking every day or some days.

SOURCE: "Figure 2. Trends in cigarette smoking among persons aged greater than or equal to 18 years, by sex—United States, 1955–1997" in "Tobacco Use—United States, 1900–1999," *Morbidity and Mortality Weekly Report*, vol. 48, no. 43, Centers for Disease Control and Prevention, November 5, 1999

to the survey). Fourteen years later, the rate of smoking among Americans had declined dramatically. In its 1999 *Survey,* SAMHSA reported that 68.2 percent of the U.S. population (186 million people) had smoked cigarettes at some time in their lives, and 25.8 percent (70.4 million) were current smokers. (See Table 5.1.) (Population estimates are based on a 1999 U.S. population of 273 million.)

In 1999, men (28.3 percent) were more likely than women (23.4 percent) to be current smokers. Additionally, whites (27.0 percent) were more likely to be current smokers than blacks (22.5 percent), Hispanics (22.6 percent), or Asians (16.6 percent). Those aged 18–25 had the highest rates of smoking.

NATIONAL HEALTH INTERVIEW SURVEY. The *National Health Interview Survey* (NHIS), conducted annually by the National Center for Health Statistics, reports findings similar to those of the NHSDA. The *National Health Interview Survey 2000* found that 23.0 percent of adults in the United States were current smokers in 2000, down slightly from 24.7 percent in 1997 (Figure 5.3) and down significantly from 42.4 percent in 1965. As revealed in the NHSDA, the NHIS found that men were more likely than women to smoke. The "current" smoker category comprised 25.6 percent of adult men and 20.6 percent of adult women. Women were more likely than men to have never smoked. (See Figure 5.4.)

Although the NHIS used different age groups than the NHSDA, results of both surveys showed that younger persons smoke at a higher rate than older persons. Figure 5.5

TABLE 5.1

Percentages reporting lifetime, past year, and past month use of cigarettes, among persons aged 12 or older, by demographic characteristics, 1999

Demographic Characteristic	TIME PERIOD		
	Lifetime	Past Year	Past Month
Total	68.2	30.1	25.8
Age			
12-17	37.1	23.5	14.9
18-25	68.9	47.5	39.7
26 or Older	72.3	28.1	24.9
Gender			
Male	73.7	33.1	28.3
Female	63.1	27.4	23.4
Hispanic Origin			
And Race			
Not Hispanic			
White Only	73.0	31.4	27.0
Black Only	56.2	25.8	22.5
American Indian or			
Alaska Native Only	73.6	42.7	36.0
Native Hawaiian			
or Other Pacific			
Islander	*	*	*
Asian Only	43.1	20.2	16.6
Multiple Race	66.4	33.4	29.8
Hispanic	55.4	28.1	22.6

*Low precision; no estimate reported.

SOURCE: "Table G.38. Percentages Reporting Lifetime, Past Year, and Past Month Use of Cigarettes Among Persons Aged 12 or Older, by Demographic Characteristics: 1999," in *Summary of Findings from the 1999 National Household Survey on Drug Abuse*, Substance Abuse and Mental Health Services Administration, Rockville, MD, 2000

shows that those aged 18–44 were more likely than other age groups to smoke. The rate of smoking was dramatically lower in the 65 and over age group.

Also, like the NHSDA, the NHIS found that the prevalence of current smoking was higher for whites (24.1 percent) than for blacks (23.1 percent) or Hispanics (17.4 percent). (See Figure 5.6.)

THE GALLUP POLL. The Gallup Organization has also observed a decline in the use of cigarettes. In 1954 nearly one-half (45 percent) of those asked indicated they had smoked within the last week; the proportion dropped to one-quarter (25 percent) in 2000. (See Table 5.2.) For more than two decades, the Gallup Organization has polled smokers on how many cigarettes they smoke each day. In 2000, 62 percent reported that they smoked less than one pack per day; 29 percent reported smoking one pack a day; and 9 percent said they smoked more than one pack a day. (See Table 5.3.)

SMOKING AND THE MILITARY. Soldiers throughout the past two centuries have smoked to calm their fears, fight fatigue, or pass the time. During the Revolutionary War, George Washington is said to have urged those on the home front, "If you can't send money, send tobacco." Popular images of soldiers in World War II, the Korean War, and Vietnam feature cigarettes prominently sticking out of

FIGURE 5.3

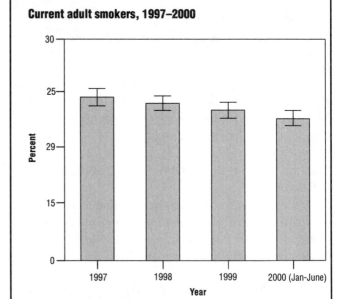

Note: Current smokers were defined as those who smoked more than 100 cigarettes in their lifetime and now smoke every day or some days. The analysis excluded people with unknown smoking status. Brackets indicated 95% confidense intervals (CI).

SOURCE: "Prevalence of current smoking adults: United States, 1997-2000" in *National Health Interview Survey, 2000,* Centers for Disease Control and Prevention, Atlanta, GA

FIGURE 5.4

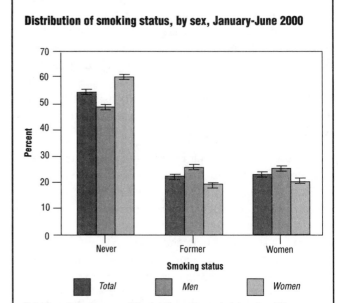

Note: Current smokers were defined as those who smoked more than 100 cigarettes in their lifetime and now smoke every day or some days. The analysis excluded 136 people with unknown smoking status. Brackets indicate 95% confidence intervals (CI).

SOURCE: "Percent distribution of smoking status among adults, by sex: United States, January–June 2000," in *National Health Interview Survey, 2000*, Centers for Disease Control and Prevention, Atlanta, GA

FIGURE 5.5

Current adult smokers by sex and age group, January-June 2000

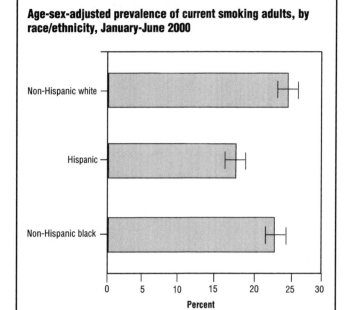

Note: Current smokers were defined as those who smoked more than 100 cigarettes in their lifetime and now smoke every day or some days. The analysis excluded 136 people with unknown smoking status. Brackets indicate 95% confidence intervals (CI).

SOURCE: "Prevalence of current smoking among adults, by sex and age group: United States, January-June 2000, " in *National Health Interview Survey, 2000,* Centers for Disease Control and Prevention, Atlanta, GA

FIGURE 5.6

Age-sex-adjusted prevalence of current smoking adults, by race/ethnicity, January-June 2000

Note: Current smokers were defined as those who smoked more than 100 cigarettes in their lifetime and now smoke every day or some days. The analysis excluded 136 people with unknown smoking status. Brackets indicate 95% confidence intervals.

SOURCE: "Age-sex-adjusted prevalence of current smoking adults, by race/ethnicity: United States, January-June 2000," in *National Health Interview Survey, 2000,* Centers for Disease Control and Prevention, Atlanta, GA

TABLE 5.2

Percentage of responses to the question "Have you, yourself, smoked any cigarettes in the past week?"

	Yes %
2000	25
1999	23
1998	28
1997	26
1996	27
1994	27
1991	28
1988	32
1985	35
1981	35
1977	38
1972	43
1969	40
1957	42
1954	45
1949	44
1944	41

2000 data based on telephone interviews with a randomly selected national sample of 1,028 persons, 18 years and older. (Error margin: plus or minus 3 percentage points)

SOURCE: Lydia Saad, "Have you, yourself, smoked any cigarettes in the past week?" in *Smoking in Restaurants Frowned on by many Americans,* Gallup News Service, Princeton, NJ, November 29, 2000

TABLE 5.3

Percentage of responses to the question "About how many cigarettes do you smoke each day?"

	Less than one pack %	One pack %	More than one pack %	No answer %	Mean number
2000	62	29	9	0	15
1999	55	35	9	1	14
1997	48	32	19	1	—
1997	48	30	21	1	—
1996	43	38	16	3	—
1994	44	38	18	0	—
1991	48	34	17	1	—
1990	51	32	14	3	—
1989	39	39	20	2	—
1988	40	38	20	2	—
1987	48	32	18	2	—
1981	38	37	24	1	—
1977	41	31	27	1	—

2000 data based on 239 respondents who smoke. (Error margin: plus or minus 7 percentage points)

SOURCE: Lydia Saad, "About how many cigarettes do you smoke each day?" in *Smoking in Restaurants Frowned on by many Americans,* Gallup News Service, Princeton, NJ, November 29, 2000

their mouths. A rest break in the military typically began with the sergeant barking, "Smoke 'em if you got 'em."

In 1975, however, authorities stopped including cigarettes in the K-rations and C-rations issued to soldiers and sailors. Effective April 8, 1994, the U.S. Department of Defense banned smoking in all military workplaces. In 1996 the Pentagon ended a subsidy that made tobacco products cheaper at military commissaries (grocery stores).

TABLE 5.4

Middle and high school students who were current users of any tobacco product—taken from National Youth Tobacco Survey, 1999

	Any tobacco*	Cigarettes	Cigars	Smokeless tobacco	Pipes	Bidis	Kreteks
Middle school							
Sex							
Male	14.2 (±2.2)†	9.6 (±1.7)	7.8 (±1.3)	4.2 (±1.3)	3.5 (±0.8)	3.1 (±0.8)	2.2 (±0.6)
Female	11.3 (±2.2)	8.9 (±1.7)	4.4 (±1.3)	1.3 (±0.5)	1.4 (±0.6)	1.8 (±0.6)	1.7 (±0.7)
Race/ethnicity							
White	11.6 (±2.3)	8.8 (±2.0)	4.9 (±1.0)	3.0 (±1.1)	2.0 (±0.6)	1.8 (±0.5)	1.7 (±0.7)
Black	14.4 (±2.7)	9.0 (±1.8)	8.9 (±2.3)	1.9 (±0.9)	2.0 (±0.9)	2.8 (±1.3)	1.7 (±0.8)
Hispanic	15.2 (±5.2)	11.0 (±4.1)	7.6 (±2.9)	2.2 (±0.9)	3.8 (±1.7)	3.5 (±1.6)	2.1 (±0.6)
Total, middle school	**12.8 (±2.0)**	**9.2 (±1.6)**	**6.1 (±1.1)**	**2.7 (±0.7)**	**2.4 (±0.5)**	**2.5 (±0.6)**	**1.9 (±0.5)**
High school							
Sex							
Male	38.1 (±3.2)	28.7 (±2.8)	20.3 (±1.9)	11.7 (±2.8)	4.2 (±0.9)	6.1 (±1.0)	6.2 (±1.1)
Female	31.4 (±3.1)	28.2 (±3.3)	10.2 (±1.6)	1.5 (±0.6)	1.4 (±0.5)	3.8 (±1.0)	5.3 (±1.5)
Race/ethnicity							
White	39.4 (±3.2)	32.9 (±3.1)	16.0 (±1.6)	8.7 (±2.1)	2.6 (±0.6)	4.4 (±0.9)	6.5 (±1.5)
Black	24.0 (±4.2)	15.9 (±3.8)	14.8 (±3.5)	2.4 (±1.3)	1.9 (±0.9)	5.8 (±2.1)	2.8 (±1.5)
Hispanic	30.7 (±4.4)	25.8 (±4.7)	13.4 (±2.9)	3.7 (±1.6)	3.8 (±1.4)	5.6 (±2.1)	5.5 (±1.9)
Total, high school	**34.8 (±2.6)**	**28.5 (±2.6)**	**15.3 (±1.4)**	**6.6 (±1.6)**	**2.8 (±0.5)**	**5.0 (±0.8)**	**5.8 (±1.2)**

* Current use of cigarettes *or* cigars *or* smokeless tobacco *or* pipes *or* bidis *or* kreteks on ≥1 of the 30 days preceding the survey.
† Ninety-five percent confidence interval.

SOURCE: "Table 4: Percentage of middle school and high school students who were current users of any tobacco product, cigarettes, cigars, smokeless tobacco, pipes, bidis, or kreteks, by sex and race/ethnicity-National Youth Tobacco Survey, 1999," in "Youth Tobacco Surveillance, United States, 1998-1999," in *Morbidity and Mortality Weekly Report*, October 13, 2000, vol. 49, no. SS-10

Cigars, Pipes, and Other Forms of Tobacco

Cigar use seems to be increasing, and cigar bars and clubs have become popular. Unlike the cigars smoked in earlier generations, which may have cost a dime or a quarter, today's "upscale" cigars cost many dollars and are often intended to display perceived success. According to the Bureau of Alcohol, Tobacco, and Firearms (ATF), U.S. smokers consumed about 3.7 billion cigars in 1998, or 37.8 cigars per male 18 years old and over.

The 1999 *National Youth Tobacco Survey,* a component of the 1998–99 *Youth Tobacco Surveillance* (*Morbidity and Mortality Weekly Report,* 2000), revealed that 6.1 percent of middle school students and 15.3 percent of high school students reported being current users of cigars. (See Table 5.4.) Male students use cigars at approximately twice the rate of female students. Of middle school students, blacks are the most likely to use cigars, followed by Hispanic students and then white students. At the high school level, however, white students are the most likely to use cigars.

Fewer students use smokeless tobacco (chewing tobacco and snuff) than cigars: 2.7 percent of middle school students and 6.6 percent of high school students. White students at both the middle school and high school level are more likely to use smokeless tobacco than black or Hispanic students.

Kreteks, or clove cigarettes, and bidi cigarettes are becoming increasingly popular among American youth. Clove cigarettes are manufactured in Indonesia and have been imported into the U.S. since 1968. They contain approximately 40 percent ground cloves and 60 percent tobacco with added clove oil. Clove cigarettes are rolled tighter than regular cigarettes and deliver, on average, twice as much tar, nicotine, and carbon monoxide as do moderate tar-containing American cigarettes. (Tars are sticky, cancer-causing substances similar to road tar. Carbon monoxide reduces the blood's ability to carry oxygen.)

Table 5.4 shows that about 1.9 percent of middle school students and 5.8 percent of high school students are current users of kreteks. These tobacco products are more popular with young males than with young females, and are used slightly more frequently by Hispanic middle school students than by black or white middle school students, and much more frequently by white and Hispanic high school students than by black high school students.

Bidis are small, strong-smelling, flavored brown cigarettes, wrapped in leaves much like cigars. They are produced in India and other Southeast Asian countries and were not widely used in the U.S. until the mid-1990s. Bidis produce approximately three times the amount of carbon monoxide and nicotine as American cigarettes, and about five times the amount of tar. Table 5.4 shows that about 2.5 percent of middle school students and 5 percent of high school students are current users of bidis. These tobacco products are more popular with black and Hispanic middle school and high school students than with white students. Additionally, they are more popular with young males than with young females.

HEALTH CONSEQUENCES

Cigarette smoke contains almost 4,000 different chemical compounds, many of which are toxic, mutagenic

TABLE 5.5

Health effects of using tobacco

System	Adverse Effects	Disease/Condition
Respiratory	Introduces carcinogens directly to lung and other respiratory tissues, and inhibits normal cleaning action of cilia	Lung, bronchus, larynx, mouth and throat cancers Emphysema Acute and chronic bronchitis Pneumonia Trigger for asthma attacks
Circulatory	Increases heart rate, elevates blood pressure, constricts blood vessels, and promotes build-up of fatty deposits in arteries	Cardiovascular disease, which includes coronary heart disease, stroke, and hypertension (chronic high blood pressure)
Digestive	Increases secretion of digestive acid, narrows blood vessels in gums, and introduces carcinogens to digestive system organs	Ulcers Periodontal disease Stomach, esophageal, and pancreatic cancer
Excretory	Contaminates urine with carcinogens	Bladder and kidney cancer
Skeletal	Causes a loss of bone density	Osteoporosis
Reproductive	Introduces carcinogens to the cervical mucus from the bloodstream	Cervical cancer

SOURCE: "Health Effects of Using Tobacco," prepared by staff of Information Plus.

(capable of increasing the frequency of mutation), and carcinogenic (cancer-causing). At least 43 carcinogens have been identified in tobacco smoke. In addition to nicotine, the most damaging substances are tar and carbon monoxide. Smoke also contains hydrogen cyanide and other chemicals that can damage the respiratory system. These substances and nicotine are absorbed into the body through the linings of the mouth, nose, throat, and lungs. About 10 seconds later, they are delivered by the bloodstream to the brain.

Tar, which adds to the flavor of cigarettes, is released by the burning tobacco. As it is inhaled, it enters the alveoli (air cells) of the lungs. There, the tar hampers the action of cilia—small, hairlike forms that clean foreign substances from the lungs—allowing the substances in cigarette smoke to accumulate.

Carbon monoxide (CO) affects the blood's ability to distribute oxygen throughout the body. CO is chemically very similar to carbon dioxide (CO_2), which bonds with the hemoglobin in blood so that the CO_2 can be carried to the lungs for elimination. Hemoglobin has two primary functions: to carry oxygen to all parts of the body and to remove excess CO_2 from the body's tissues. CO bonds to hemoglobin more tightly than CO_2 and also leaves the body more slowly, which allows CO to build up in the hemoglobin, in turn reducing the amount of oxygen the blood can carry. Lack of adequate oxygen is damaging to most of the body's organs, including the heart and brain.

Smokeless tobacco, which includes chewing tobacco and snuff, also creates health hazards for its users. In 1979 the yearly *Report of the Surgeon General* noted that smokeless tobacco was associated with oral cancers; in the 1986 *Report,* the surgeon general concluded that it

was a cause of these diseases. The nicotine in smokeless tobacco is absorbed into the bloodstream through the lining of the mouth and has been linked to periodontal (gum) disease, as well as cancers of the lip, gum, and mouth.

Diseases and Conditions Linked to Tobacco Use

Medical research has associated smoking with cancer, heart and circulatory disease, fetal growth retardation, and low-birthweight babies. The 1983 *Report of the Surgeon General* linked cigarette smoking to cerebrovascular disease (stroke) and associated it with cancer of the uterine cervix. Two 1992 studies showed that people who smoke double their risk of forming cataracts, the leading cause of blindness. Recent research links smoking to unsuccessful pregnancies, increased infant mortality, and peptic ulcer disease. It is also a contributor to cancer of the bladder, pancreas, and kidney, and is associated with cancer of the stomach. Table 5.5 summarizes some of the links between tobacco use and the risks of developing certain health problems.

In 1998 the National Cancer Institute noted the following about cigar smoking: (1) Cigars contain most of the same cancer-causing chemicals that are found in cigarettes. (2) Regular cigar smoking causes cancer of the lungs, mouth, larynx (voice box), esophagus (food tube), and probably cancer of the pancreas. (3) Cigar smokers have four to ten times the risk of dying of cancers of the larynx, mouth, or esophagus than nonsmokers.

Reporting on a study funded by the National Cancer Institute, lead author Carlos Iribarren, in "Effect of Cigar Smoking on the Risk of Cardiovascular Disease, Chronic Obstructive Pulmonary Disease, and Cancer in Men" (*New England Journal of Medicine,* June 1999), reported that cigar smokers are twice as likely as nonsmokers to develop

TABLE 5.6

Harmful interactions between smoking and medications

Medications	Effects
Acetaminophen (over-the-counter non-aspirin pain relievers)	Decreases the effect of acetaminophen
Antidepressants	Decreases the antidepressant effect
Some sedatives and tranquilizers	May reduce their relaxation effect
Estrogens and oral contraceptives	May decrease estrogen effect; increases risk of heart and blood-vessel disease
Ulcer medications	May slow ulcer healing
Insulin	Reduces the effect of insulin

SOURCE: Prepared by staff of Information Plus

TABLE 5.7

Selected Reports on Smoking and Health by the U.S. Surgeon General

Year	Topic
1964	Smoking and Health
1967	The Health Consequences of Smoking
1980	The Health Consequences of Smoking for Women
1982	The Health Consequences of Smoking—Cancer
1983	The Health Consequences of Smoking—Cardiovascular Disease
1984	The Health Consequences of Smoking—Chronic Obstructive Lung Disease
1985	The Health Consequences of Smoking—Cancer and Chronic Lung Disease in the Workplace
1986	The Health Consequences of Involuntary Smoking
1988	The Health Consequences of Smoking—Nicotine Addiction
1990	The Health Benefits of Smoking Cessation
1994	Preventing Tobacco Use Among Young People
1998	Tobacco Use Among U.S. Racial/Ethnic Minority Groups
2000	Reducing Tobacco Use
2001	Women and Smoking

SOURCE: Prepared by staff of Information Plus

cancer of the mouth, throat, and lungs. Cigar smokers are also more likely to develop heart disease or chronic pulmonary disease. Iribarren observed, "Many people still believe it is safe to smoke cigars. Our research shows that there are serious health consequences for cigar smokers."

In an editorial accompanying the article cited above, Surgeon General Dr. David Satcher said, "Restrictions on the sale of cigars (through the setting of excise rates, for example) ought to be at least as stringent as those currently applied to other tobacco products." Dr. Satcher has urged the Federal Trade Commission to require warning labels on cigars, like those on cigarettes.

Analyzing the health problems of smokers, the Public Health Service estimated that smokers annually miss 81 million days of work and spend 145 million days sick in bed. Compared with nonsmokers, smokers had, per year:

• 11 million more cases of chronic illnesses.

• 280,000 additional cases of heart disease.

• 1 million more cases each of chronic bronchitis, emphysema, and peptic ulcer.

• 1.8 million more cases of sinus problems.

The National Institute on Drug Abuse, in *Nicotine Addiction* (U.S. Department of Health and Human Services, July 1998), found smoking responsible for approximately 7 percent of total U.S. health care costs, or about $50 billion each year. However, this amount is only about half the total cost of smoking on society, because it does not include perinatal care for low-birthweight infants of mothers who smoke; medical care costs associated with diseases caused by secondhand smoke; and burn care from smoking-related fires. Including these costs, the total financial burden of smoking is estimated at more than $100 billion a year.

Premature Aging

Smoking cigarettes contributes to premature aging in a variety of ways. Results of research over two decades

show that smoking enhances facial aging and skin wrinkling (M.F. Demierre et al., "Public Knowledge, Awareness, and Perceptions of the Association between Skin Aging and Smoking," *Journal of the American Academy of Dermatology,* 1999). Additionally, smoking has been associated with a decline in fitness in women, which makes a female smoker, especially a heavy smoker, act several years older than she really is (Heidi D. Nelson et al., "Smoking, Alcohol, and Neuromuscular and Physical Function of Older Women," *Journal of the American Medical Association,* 1994).

Interactions with Other Drugs

Smoking can have adverse effects when combined with over-the-counter (OTC) and prescription medications that a smoker may be taking. In many cases tobacco smoking reduces the effectiveness of medications, such as pain relievers (acetaminophen), antidepressants, tranquilizers, sedatives, ulcer medications, and insulin. With estrogen and oral contraceptives, tobacco smoking may increase the risk of heart and blood vessel disease. Table 5.6 lists some of the adverse effects of combining smoking and medications.

Smoking and Public Health

Nearly 40 years ago, the U.S. government first officially recognized the negative health consequences of smoking. And earlier, in the 1920s, a study had found that men who smoked two or more packs of cigarettes per day were 22 times more likely than nonsmokers to die of lung cancer. At the time, these results surprised researchers and medical authorities alike.

In 1964 the Advisory Committee to the Surgeon General released a groundbreaking survey of studies on tobac-

TABLE 5.8

Leading causes of death and numbers of deaths, 1980 and 1998

Rank order	1980 Cause of death	Deaths	1998 Cause of death	Deaths
	All causes	1,989,841	All causes	2,337,256
1	Diseases of heart	761,085	Diseases of heart	724,859
2	Malignant neoplasms	416,509	Malignant neoplasms	541,532
3	Cerebrovascular diseases	170,225	Cerebrovascular diseases	158,448
4	Unintentional injuries	105,718	Chronic obstructive pulmonary diseases	112,584
5	Chronic obstructive pulmonary diseases	56,050	Unintentional injuries	97,835
6	Pneumonia and influenza	54,619	Pneumonia and influenza	91,871
7	Diabetes mellitus	34,851	Diabetes mellitus	64,751
8	Chronic liver disease and cirrhosis	30,583	Suicide	30,575
9	Atherosclerosis	29,449	Nephritis, nephrotic syndrome, and nephrosis	26,182
10	Suicide	26,869	Chronic liver disease and cirrhosis	25,192

SOURCE: "Table 32: Leading causes of death and numbers of deaths, according to sex, detailed race, and Hispanic origin: United States, 1980 and 1998" in *Health, United States, 2000,* National Center for Health Statistics. Hyattsville, MD, 2000

co use. In *Smoking and Health: Report of the Advisory Committee to the Surgeon General of the Public Health Service,* the surgeon general reported that cigarette smoking increased overall mortality in men and caused lung and laryngeal cancer, as well as chronic bronchitis. The report concluded, "Cigarette smoking is a health hazard of sufficient importance in the United States to warrant appropriate remedial action," but what action should be taken was left unspecified at that time.

Later surgeons general issued additional reports on the health effects of smoking and the dangers to nonsmokers of "passive" or "secondhand" smoke. In addition to general health concerns, the reports have addressed specific health consequences and populations. Table 5.7 shows a listing of selected reports and the years in which they were published. The later reports concluded that smoking increased the morbidity (proportion of diseased persons in a particular population) and mortality (proportion of deaths in a particular population) of both men and women.

In 1965 Congress passed the Federal Cigarette Labeling and Advertising Act (PL 89-92), which required the following health warning on all cigarette packages: "Caution: Cigarette smoking may be hazardous to your health." The Public Health Cigarette Smoking Act of 1969 (PL 91-222; passed in 1970) strengthened the warning to read: "Warning: The Surgeon General has determined that cigarette smoking is dangerous to your health." Still later acts resulted in four different health warnings to be used in rotation.

The *Morbidity and Mortality Weekly Report* (April 1999) included "recognition of tobacco use as a health hazard" as one of the country's ten greatest public health achievements of the 20th century, along with vaccination, control of infectious diseases, safer and healthier food, healthier mothers and babies, family planning, safer workplaces, motor-vehicle safety, decline in deaths from coronary heart disease and stroke, and fluoridation of drinking water. These ten accomplishments were chosen based on their contributions to prevention and their impact on illness, disability, and death in the United States.

Deaths Attributed to Tobacco Use

According to the CDC, "Cigarette smoking is the single most preventable cause of premature death in the United States." The CDC estimates that more than 430,000 Americans die each year from the effects of cigarette smoking. About one in every five deaths (20 percent) is smoking-related. Nationwide, smoking kills more people each year than alcohol, drug abuse, car crashes, murders, suicides, fires, and AIDS combined.

In 1998 diseases linked to smoking accounted for four of the top five leading causes of death in the United States. (See Table 5.8.) Nearly 725,000 persons died of various heart diseases in 1998 (down from about 761,000 in 1980). Cerebrovascular disease (stroke) claimed 158,448 lives. About 112,500 died of chronic obstructive pulmonary diseases, including chronic bronchitis, asthma, and emphysema. For 2001, the American Cancer Society estimated that 157,400 people would die of lung cancer. While not all lung cancer deaths are directly attributable to smoking, a large proportion of them are.

Lung cancer is the leading cause of cancer mortality in both men and women in the United States. It has been the leading cause of cancer deaths among men since the early 1950s and, in 1987, surpassed breast cancer to become the leading cause of cancer deaths in women. The American Cancer Society estimated that, in 2001, 90,100 men and 67,300 women would die from lung and bronchus cancer in the United States, accounting for 28 percent of all cancer deaths. Male rates have decreased by 7 percent since 1980, while female rates have increased by 53 percent. From 1990 to 1997, mortality from lung cancer declined significantly among men (-1.7 percent per year), while rates for women increased significantly (0.9 percent per year).

ENVIRONMENTAL TOBACCO SMOKE

Smoking is also a health hazard for nonsmokers who live or work with smokers. A 1992 report by the U.S. Environmental Protection Agency (EPA), *Respiratory Health Effects of Passive Smoking: Lung Cancer and Other Disorders* (Washington, D.C.), concluded that the "widespread exposure to environmental tobacco smoke (ETS) in the United States presents a serious and substantial public health impact." Some of its findings were:

- In adults, ETS is a human lung carcinogen, responsible for approximately 3,000 lung cancer deaths annually in U.S. nonsmokers. A person living with a spouse who smokes has a 20–50 percent increased risk of developing lung cancer.

- In children, ETS exposure is causally associated with an increased risk of lower respiratory tract infections such as bronchitis and pneumonia. During the first year of life, infants whose mothers smoke experience 32 percent more bronchitis and pneumonia. Exposure to smoke may also account for 10–35 percent of chronic middle-ear problems in children.

- Children's exposure to ETS is causally associated with increased prevalence of fluid in the middle ear, symptoms of upper respiratory tract irritation, and a small but significant reduction in lung function.

- Childhood ETS exposure is causally associated with additional and increased severity of symptoms in children with asthma. This report estimated that exposure to ETS has worsened the conditions of two hundred thousand to one million asthmatic children.

- ETS exposure is a risk factor for new cases of asthma in children who have not previously displayed symptoms.

According to the EPA, secondhand cigarette smoke kills 53,000 nonsmokers a year, including 37,000 from heart disease. Responding to this study, tobacco company R.J. Reynolds insisted that nonsmokers are normally exposed to very little secondhand smoke. The Tobacco Institute, which represented the tobacco industry until January 1999, contended that the study did not include other factors, such as diet and medical care, which might affect the likelihood of the children's diseases. Spokesmen for the tobacco industry complained that the EPA report has led to hundreds of government and private efforts to ban or restrict indoor smoking, which could result in economic harm to the industry.

Elizabeth T. H. Fontham et al., in "Environmental Tobacco Smoke and Lung Cancer in Nonsmoking Women" (*Journal of the American Medical Association,* June 8, 1994), found that tobacco use by a spouse increased the likelihood of cancer by 30 percent among the women studied. If the spouse was a heavy smoker, the chances increased to 80 percent.

If, in addition to exposure at home, the woman was also exposed at work, she was 39 percent more likely to develop cancer, and if she was also exposed in social settings, it rose to 50 percent. The likelihood of cancer increased the longer the woman remained in such situations. On the other hand, the researchers found no significant association between cancer and exposure from parents or other smokers when the woman was a child.

In 1997 the results of a large-scale study of 32,046 U.S. female nurses, examining the relationship between passive smoking and the risk of coronary heart disease (CHD), were published. Ichiro Kawachi and colleagues ("A Prospective Study of Passive Smoking and Coronary Heart Disease," *Circulation,* May 20, 1997) found that occasional exposure to environmental tobacco smoke increased a female nonsmoker's chance of CHD by 58 percent, and regular exposure to passive smoke increased her risk by 90 percent.

Robert M. Davis, M.D., in "Exposure to Environmental Tobacco Smoke" (*Journal of the American Medical Association,* December 1998), observed that nearly everyone in the United States is at some risk of harm from secondhand smoke. Not surprisingly, those at greater risk of harm are those who live with smokers and those who work where smoking is allowed. The longer the time spent in a smoking environment and the greater the concentration of ETS in that airspace, the more risk for harm. The concentration of ETS is affected by the size of the space, the number of people smoking there, and the ventilation rate. In 2000 the CDC reported that the proportion of survey respondents from 17 states and the District of Columbia who reported a smoke-free policy at their indoor workplace ranged from 61.3 percent to 82 percent.

Some people, however, are more susceptible to harm from ETS exposure because of their age or health status. Because their lungs and other respiratory tissue are still developing, infants and children exposed to ETS are more likely to develop asthma, bronchitis, pneumonia, and middle ear disease. People with certain chronic conditions, such as asthma, allergies, and chronic lung disease, may be more susceptible to the harmful effects of secondhand smoke.

A 1999 Gallup Poll asked respondents, "In general, how harmful do you feel secondhand smoke is to adults?" Forty-three percent thought secondhand smoke was very dangerous, and another 39 percent believed it was somewhat dangerous. In summary, eight of ten Americans say secondhand smoke is harmful. (See Table 5.9.)

THE MOVE TO BAN SMOKING

During the 1980s and 1990s, many Americans became more concerned about the effects of smoking. In some areas, smokers have become virtual outcasts, unable to smoke where they work, at parties, in many restaurants, in airplanes, and sometimes even when they visit their best friends. Today, virtually every state has passed laws banning cigarettes in at least some public facilities.

In 1991 the federal government instituted its largest antismoking campaign, intended to prevent 1.2 million smoking-related deaths. The goal of the seven-year program was to help 5.5 million adults stop smoking, prevent 2 million youths from starting, and reduce smokers to 15 percent of the population. One of the national health objectives for the year 2010 (*Healthy People 2010*, 2nd ed., U.S. Department of Health and Human Services, Washington, D.C., November 2000) is to reduce the prevalence of cigarette smoking among adults to no more than 12 percent. As noted in "Cigarette Smoking among Adults—United States, 1998" (*Morbidity and Mortality Weekly Report*, October 6, 2000) accomplishing this goal will require aggressive public health efforts to implement comprehensive tobacco-control programs nationwide.

Numerous health organizations, citing risks to nonsmokers, advocate the banning of smoking in all public places. Action on Smoking and Health (ASH), based in Washington, D.C., was formed to educate the public on the hazards of smoking and to protect the rights of nonsmokers. ASH supports state legislation that restricts smoking.

Global Efforts to Reduce Tobacco Use

In May 1999 member states of the World Health Organization (WHO) approved a resolution to initiate a Framework Convention on Tobacco Control, possibly leading to an international treaty for global cooperation in reducing the negative health consequences of tobacco use. Currently, smoking kills one in ten adults worldwide. Without coordinated action to reduce tobacco use, this number is expected to increase significantly by 2020 as a result of free trade in tobacco products, worldwide advertising and promotion, and increased proportions of young people using tobacco.

STOPPING SMOKING

The CDC, in "Cigarette Smoking among Adults—United States, 1998" (*Morbidity and Mortality Weekly Report*, October 6, 2000), estimated that in 1998 there were 47.2 million current smokers and 44.8 million former smokers in the United States. When asked if, during the past 12 months, they had stopped smoking for one day or longer because they were trying to quit, nearly 40 percent said they had.

In a 2000 Gallup Poll, 239 smokers were asked if they would like to give up smoking. Eighty-two percent answered "yes." This figure is up from 76 percent in 1999 and from 66 percent in 1977 (Lydia Saad, "Smoking in

TABLE 5.9

Percentage of responses to the question "In general, how harmful do you feel second-hand smoke is to adults?"

	Very harmful	Somewhat harmful	Not too harmful	Not at all harmful	Depends (vol.)	No opinion
1999	43%	39%	11%	5%	1%	1%
1997	55	29	9	5	*	2
1996	48	36	9	5	*	2
1994	36	42	12	6	1	3

1999 data based on telephone interviews with a randomly selected national sample of 1,039 persons,18 years and older. (Error margin: plus or minus 3 percentage points) (vol.) means volunteered response.
*means less than 0.5%.

SOURCE: David W. Moore, "In general, how harmful do you feel second-hand smoke is to adults—Very harmful, somewhat harmful, not too harmful, or not at all harmful?" in *Nine of Ten Americans View Smoking as Harmful,* Gallup News Service, Princeton, NJ, October 7, 1999

Restaurants Frowned on by Many Americans," *Gallup News Service,* November 29, 2000). The 1997 National Health Interview Survey found that about 22.8 percent of adults in the U.S. (25.1 million men and 19.2 million women) were former smokers.

Depression a Factor?

Studies released in 1990, performed at the New York State Psychiatric Institute and the CDC in Atlanta, found that smokers are more likely than nonsmokers to have suffered from depression. Researchers at the Psychiatric Institute found that smokers were more than twice as likely as nonsmokers to have a history of serious depression.

Severe depression seems to make it very difficult for some smokers to quit. In a follow-up survey of smokers nine years after the original survey, results of the CDC study of depressive symptoms and smoking revealed that only 10 percent of depressed smokers had quit, compared with 18 percent of smokers who were not depressed. Many observers believe this supports a widely held theory that smokers, like drug addicts, are smoking to alleviate depression and that the effects of nicotine on the brain offer the depressed smoker some relief.

The Benefits of Stopping

The Health Benefits of Smoking Cessation: A Report of the Surgeon General, 1990 (Washington, D.C., 1990) reported that quitting offers major and immediate health benefits for both sexes and for all ages. This first comprehensive report on the benefits of quitting showed that many of the ill effects of smoking can be reversed. For persons who quit smoking, the health risks decline steadily for 15 years, when the risk of death returns to the level of persons who have never smoked. Male ex-smokers between ages 35 and 49 add an average of five years to

FIGURE 5.7

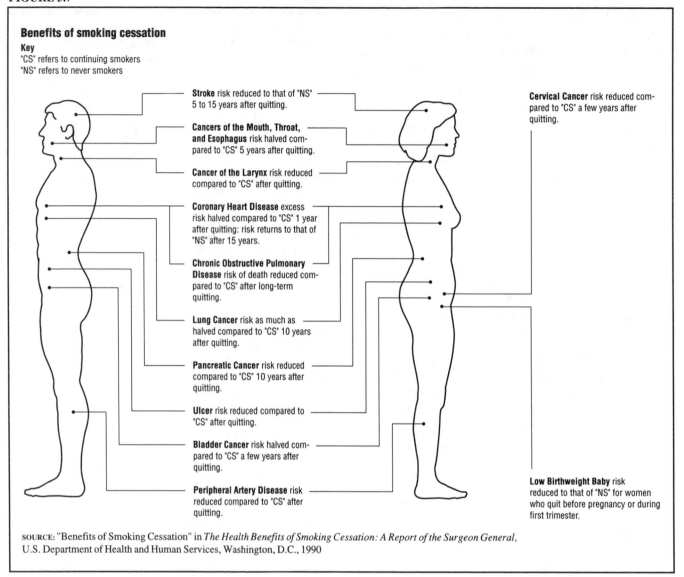

Benefits of smoking cessation

Key
"CS" refers to continuing smokers
"NS" refers to never smokers

Stroke risk reduced to that of "NS" 5 to 15 years after quitting.

Cancers of the Mouth, Throat, and Esophagus risk halved compared to "CS" 5 years after quitting.

Cancer of the Larynx risk reduced compared to "CS" after quitting.

Coronary Heart Disease excess risk halved compared to "CS" 1 year after quitting: risk returns to that of "NS" after 15 years.

Chronic Obstructive Pulmonary Disease risk of death reduced compared to "CS" after long-term quitting.

Lung Cancer risk as much as halved compared to "CS" 10 years after quitting.

Pancreatic Cancer risk reduced compared to "CS" 10 years after quitting.

Ulcer risk reduced compared to "CS" after quitting.

Bladder Cancer risk halved compared to "CS" a few years after quitting.

Peripheral Artery Disease risk reduced compared to "CS" after quitting.

Cervical Cancer risk reduced compared to "CS" a few years after quitting.

Low Birthweight Baby risk reduced to that of "NS" for women who quit before pregnancy or during first trimester.

SOURCE: "Benefits of Smoking Cessation" in *The Health Benefits of Smoking Cessation: A Report of the Surgeon General,* U.S. Department of Health and Human Services, Washington, D.C., 1990

their lives, while females add three years. Figure 5.7 illustrates the benefits of smoking cessation.

After just one year of nonsmoking, the smoking-related excess risk of heart disease is reduced by half. The risk of stroke returns to that of someone who has never smoked within 5–15 years. After 10 years, those who quit smoking reduce their risk of lung cancer by as much as 50 percent, compared with continuing smokers, and the risk continues to decline with additional smoke-free years. Ex-smokers are also less likely to die of chronic lung diseases, such as emphysema. Overall, former smokers improve their health status, lower their number of sick days, and reduce their rates of bronchitis and pneumonia.

Results of a recent study published in the *British Medical Journal* ("Smoking, Smoking Cessation, and Lung Cancer in the UK since 1950: Combination of National Statistics with Two Case-Control Studies," August 5, 2000) revealed the extent to which smoking

cessation lowers lung cancer risk. For men who stopped smoking at ages 60, 50, 40, and 30, the cumulative risks of lung cancer by age 75 were 10 percent, 6 percent, 3 percent, and 2 percent respectively. This means that persons who quit smoking in middle age or prior to middle age avoid more than 90 percent of the lung cancer risk attributable to tobacco.

Another recent study ("Effects of Multiple Attempts to Quit Smoking and Relapses to Smoking on Pulmonary Function," *Journal of Clinical Epidemiology,* December 1998) investigated whether short periods of quitting were beneficial to smokers' health. Results revealed that those who made several attempts to quit smoking had less loss of lung function than those who continued to smoke. Therefore, even intermittent lapses in smoking are beneficial.

UNIQUE BENEFITS FOR WOMEN. The 1990 *Report of the Surgeon General* noted that 5 percent of deaths among newborns could be prevented if all women quit smoking during their pregnancies. Smoking by pregnant women is linked to an increased risk of miscarriage, stillbirth, premature delivery, and sudden infant death syndrome (SIDS), and is a cause of low birthweight in infants. A woman who stops smoking, either before she becomes pregnant or during her first trimester (three months) of pregnancy, significantly reduces her chances of having a low-birthweight baby. Research has found that it takes smokers longer to get pregnant than nonsmokers, but that women who quit are as likely to get pregnant as those who have never smoked.

In 1999 the *National Household Survey on Drug Abuse* found that, of the sample of females aged 15–44 who were surveyed, 31 percent used tobacco in the month prior to the survey. Of the pregnant women in this sample, 17 percent had smoked cigarettes in the prior month, 0.9 percent smoked cigars, and 0.1 percent smoked pipes. (See Table 5.10.)

Complaints about Quitting

One of the most common complaints among ex-smokers is that they gain weight when they stop smoking. In fact, nearly 80 percent of those who quit do gain weight, compared with 56 percent of continuing smokers. However, the 1990 *Report of the Surgeon General* found that the average weight gain among ex-smokers was just five pounds, and that less than 4 percent of those who quit gain more than 20 pounds after quitting. Also, people who stop smoking often start exercise programs to improve their health. Exercise not only keeps them away from cigarettes, but also prevents or minimizes weight gain. Continuing to smoke is far more dangerous to one's health than gaining a few extra pounds.

Another side effect of smoking cessation is nicotine withdrawal. Short-term consequences include anxiety, irritability, frustration, anger, difficulty concentrating, and restlessness. Possible long-term consequences are urges to smoke and increased appetite. Nicotine withdrawal symptoms peak in the first few days after quitting and subside during the following weeks. Improved self-esteem and an increased sense of control often accompany long-term abstinence.

Ways to Stop Smoking

Nicotine-replacement treatments are proving effective for many smokers. The nicotine patch is an adhesive resembling a bandage that can be attached to the skin. It

TABLE 5.10

Percentages of females (aged 15 to 44) reporting past month use of tobacco, by pregnancy status, 1999

	Total[1]	PREGNANCY STATUS	
		Pregnant	Not Pregnant
Any tobacco[2]	31.0	17.6	31.6
Cigarettes	30.0	17.0	30.5
Smokeless tobacco	0.5	0.4	0.5
Cigars	2.8	0.9	2.9
Pipes	0.3	0.1	0.3

[1] Estimates in the total column are for all females aged 15 to 44, including those with missing pregnancy status.

[2] Use of any tobacco product indicates using at least once cigarettes, smokeless tobacco (i.e., chewing tobacco or snuff), cigars, or pipe tobacco.

SOURCE: "Table G.27. Percentages Reporting Past Month Use of Tobacco and Alcohol Among Females Aged 15 to 44, by Pregnancy Status: 1999," in *Summary of Findings from the 1999 National Household Survey on Drug Abuse*, Substance Abuse and Mental Health Services Administration, Rockville, MD, 2000

contains about 30 milligrams of nicotine. The nicotine is absorbed through the skin at a rate of about one milligram per hour, thus allowing a smoker to cope with the withdrawal symptoms that discourage many smokers trying to stop. Two types of transdermal nicotine patches have become over-the-counter products.

Nicotine gum is also available over the counter. The nicotine in the gum is absorbed through the mouth and throat. Users can chew the gum at a rate that minimizes their withdrawal symptoms and can gradually reduce their intake of nicotine. In 1996 a nicotine nasal spray and, in 1998, a nicotine inhaler became available by prescription. Estimates based on FDA (Food and Drug Administration) and pharmaceutical industry data indicate that more than one million individuals have used these treatments successfully.

Non-nicotine therapies are also being developed for the relief of nicotine withdrawal symptoms. Zyban, an antidepressant, was approved as a pharmacological treatment for the addiction of nicotine in 1997. It is widely used and doubles quitting rates. Behavioral treatments, such as formal smoking-cessation programs, are successful for some smokers who want to quit. Behavioral methods are designed to create an aversion to smoking, develop self-monitoring of smoking behavior, and establish alternative coping responses.

CHAPTER 6
ALCOHOL, TOBACCO, AND YOUTH

SURVEYS OF STUDENT DRUG, ALCOHOL, AND TOBACCO USE

Three surveys provide comprehensive coverage of the use of drugs, alcohol, and tobacco by American youth, and their attitudes toward using these substances. The surveys presented in this chapter include:

• *Monitoring the Future* (Rockville, Maryland, 2001). Prepared by the Institute for Social Research of the University of Michigan for the National Institute on Drug Abuse (NIDA), the annual *Monitoring the Future* survey tracks the use of drugs, including alcohol and tobacco, among students in the eighth, tenth, and twelfth grades. It is considered an authoritative source on drug use among students. In all, approximately 50,000 students in about 420 public and private secondary schools complete the self-administered questionnaire every year.

• *PRIDE Questionnaire Report: 1999–2000 National Summary, Grades 6 through 12* (Bowling Green, Kentucky, 2000). In addition to asking about use, the annual PRIDE (Parents' Resource Institute for Drug Education) survey also questions students about when and where they use the substances, how hard the drugs are to obtain, and what other risk behaviors are present, such as violence, poor grades, and weapon carrying.

• *Youth Risk Behavior Surveillance—United States 1999* (Centers for Disease Control and Prevention, Atlanta, Georgia, 2000). In addition to asking about the use of alcohol, tobacco, and illicit drugs, the biennial *Youth Risk Behavior Surveillance* reports on other types of risk behavior, such as violence at school, weapon carrying, and sexual conduct.

All of these surveys collect self-reported data—that is, behavior reported by the students themselves. Therefore, the data should be used as indicators primarily to identify prevalence trends and patterns of use.

All the surveys listed above agree that alcohol, because of its easy availability and social acceptance, is both the most widely used drug and the greatest threat to American youth. However, Dr. Lloyd Johnston, director of the *Monitoring the Future* survey, has also expressed considerable concern over the number of students using tobacco. Commenting on the slight decline in smoking among secondary school students since 1997, Johnston and his collaborators observed that tobacco settlement efforts, in addition to proposals in Congress and by the president to bring about tobacco control legislation, stimulated a great deal of publicity about smoking and its adverse consequences. Johnston speculates that the publicity may have helped change young people's views about smoking. "If that is the case," Johnston observed, "then there is a real question about whether teen smoking will continue to decline in the absence of an intense public debate." He warns that

> The implications of adolescent smoking for the country's future rates of disease, early death, disrupted families, worker productivity, and health-care costs cannot be overestimated. . . . Cigarettes will kill far more of today's children than all other drugs combined, including alcohol. But because these consequences do not emerge for a few decades, we seem to be much less concerned about them. If cigarette smoking killed quickly, like drunk driving does, the country would be treating the current rates of adolescent smoking as an extreme emergency.

ATTITUDES TOWARD USING ALCOHOL AND TOBACCO

How young people feel about the risk of using alcohol and tobacco and how much they disapprove of their use influence how likely they are to try these substances. Although a significant percentage of students do not disapprove of, or consider very harmful, some alcohol and tobacco use, the majority do disapprove and consider

TABLE 6.1

Trends in harmfulness of drugs as perceived by eighth, tenth, and twelfth graders, 1991–2000

Percentage saying "great risk"[a]

How much do you think people risk harming themselves (physically or in other ways), if they….	8th Grade											10th Grade											12th Grade										
	1991	1992	1993	1994	1995	1996	1997	1998	1999	2000	'99-'00 change	1991	1992	1993	1994	1995	1996	1997	1998	1999	2000	'99-'00 change	Class of 1991	Class of 1992	Class of 1993	Class of 1994	Class of 1995	Class of 1996	Class of 1997	Class of 1998	Class of 1999	Class of 2000	'99-'00 change
Try one or two drinks of an alcoholic beverage (beer, wine, liquor)	11.0	12.1	12.4	11.6	11.6	11.8	10.4	12.1	11.6	11.9	+0.3	9.0	10.1	10.9	9.4	9.3	8.9	9.0	10.1	10.5	9.6	-1.0	9.1	8.6	8.2	7.6	5.9	7.3	6.7	8.0	8.3	6.4	-1.9s
Take one or two drinks nearly every day	31.8	32.4	32.6	29.9	30.5	28.6	29.1	30.3	29.7	30.4	+0.7	36.1	32.6	35.9	32.5	31.7	31.2	31.8	31.9	32.9	32.3	-0.6	32.7	30.6	28.2	27.0	24.8	25.1	24.8	24.3	21.8	21.7	-0.1
Have five or more drinks once or twice each weekend	59.1	58.0	57.7	54.7	54.1	51.8	55.6	56.0	55.3	55.9	+0.6	54.7	55.9	54.9	52.9	52.0	50.9	51.8	52.5	51.9	51.0	-0.9	48.6	49.0	48.3	46.5	45.2	49.5	43.0	42.8	43.1	42.7	-0.4
Smoke one or more packs of cigarettes per day[b]	51.6	50.8	52.7	49.8	50.4	52.6	54.3	54.8	54.8	58.8	+4.0ss	60.3	59.3	60.7	59.0	57.0	57.9	59.9	61.9	62.7	65.9	+3.3s	69.4	69.2	69.5	67.6	65.6	68.2	68.7	70.8	70.8	73.1	+2.3
Use smokeless tobacco regularly	35.1	35.1	36.9	35.5	33.5	34.0	35.2	36.5	37.1	39.0	+1.9	40.3	39.6	44.2	42.2	38.2	41.0	42.2	42.8	44.2	46.7	+2.5s	37.4	35.5	38.9	36.6	33.2	37.4	38.6	40.9	41.1	42.2	+1.1
Approx. N (in thousands) =	*17.4*	*18.7*	*18.4*	*17.4*	*17.5*	*17.9*	*18.8*	*18.1*	*16.7*	*16.7*		*14.7*	*14.8*	*15.3*	*15.9*	*17.0*	*15.7*	*15.6*	*15.0*	*13.6*	*14.3*		*2.5*	*2.7*	*2.8*	*2.6*	*2.6*	*2.4*	*2.6*	*2.6*	*2.3*	*2.1*	

Notes: Level of significance of difference between the two most recent classes: s = .05, ss = .01, sss = .001. '—' indicates data not available. Any apparent inconsistency between the change estimate and the prevalence of use estimates for the two most recent classes is due to rounding error.

a Answer alternatives were: (1) No risk, (2) Slight risk, (3) Moderate risk, (4) Great risk, and (5) Can't say, drug unfamiliar.

b Beginning in 1999 for eighth and tenth graders, data based on two-thirds of N indicated due to changes in questionnaire forms.

SOURCE: Lloyd D. Johnston, Patrick M. O'Malley, and Jerald G. Bachman, "Trends in Harmfulness of Drugs as Perceived by Eighth and Tenth Graders, 1991–2000" and "Long-Term Trends in Harmfulness of Drugs as Perceived by Twelfth Graders" in *Monitoring the Future: National Results on Adolescent Drug Use—Overview of Key Findings, 2000*, Institute for Social Research, University of Michigan, and the National Institute on Drug Abuse, U.S. Department of Health and Human Services, 2001

TABLE 6.2

PRIDE Survey: Students' beliefs about the dangers of tobacco and alcohol

Do you feel that using cigarettes is harmful to your health?

Grade Level	N of Valid	N of Miss	No Harm	Some Harm	Harmful	Very Harmful
6th	18417	285	3.4	8.9	32.9	54.8
7th	16852	204	3.9	13.1	35.9	47.1
8th	23184	301	3.8	14.8	34.7	46.8
9th	15749	206	4.3	16.4	35.2	44.1
10th	16562	194	3.6	16.4	35.7	44.3
11th	10570	114	3.5	17.1	35.4	44.0
12th	11576	104	3.5	16.0	34.5	46.0
JrHs	58453	790	3.7	12.5	34.5	49.4
SrHS	54457	618	3.8	16.5	35.2	44.5
Total	112910	1408	3.7	14.4	34.8	47.1

Do you feel that using beer is harmful to your health?

Grade Level	N of Valid	N of Miss	No Harm	Some Harm	Harmful	Very Harmful
6th	18328	374	6.8	25.6	30.1	37.6
7th	16750	306	8.4	30.1	29.7	31.9
8th	23083	402	9.0	31.8	29.5	29.7
9th	15706	249	11.2	35.3	27.7	25.8
10th	16495	261	10.7	37.0	27.8	24.5
11th	10552	132	10.7	38.4	27.7	23.1
12th	11534	146	10.8	39.4	26.5	23.2
JrHs	58161	1082	8.1	29.3	29.7	32.8
SrHS	54287	788	10.9	37.3	27.5	24.3
Total	112448	1870	9.5	33.2	28.6	28.7

Do you feel that using wine coolers is harmful to your health?

Grade Level	N of Valid	N of Miss	No Harm	Some Harm	Harmful	Very Harmful
6th	18155	547	15.0	33.4	25.3	26.2
7th	16698	358	18.9	37.6	22.6	20.9
8th	23024	461	21.8	38.0	20.9	19.3
9th	15653	302	25.2	40.1	18.4	16.4
10th	16484	272	23.2	41.3	19.0	16.4
11th	10546	138	22.9	42.5	18.5	16.1
12th	11532	148	22.2	43.1	18.5	16.2
JrHs	57877	1366	18.8	36.4	22.8	22.0
SrHS	54215	860	23.5	41.6	18.6	16.3
Total	112092	2226	21.1	38.9	20.8	19.2

Do you feel that using liquor is harmful to your health?

Grade Level	N of Valid	N of Miss	No Harm	Some Harm	Harmful	Very Harmful
6th	18221	481	4.5	14.4	31.8	49.2
7th	16703	353	5.2	18.8	33.8	42.2
8th	23047	438	6.1	20.8	33.5	39.5
9th	15668	287	8.2	25.0	32.3	34.6
10th	16493	263	8.0	26.2	32.4	33.5
11th	10532	152	8.3	28.5	31.9	31.2
12th	11522	158	8.0	30.4	31.7	29.9
JrHs	57971	1272	5.4	18.2	33.1	43.4
SrHS	54215	860	8.1	27.2	32.1	32.6
Total	112186	2132	6.7	22.5	32.6	38.2

SOURCE: "6.30–Do you feel that using cigarettes is harmful to your health?" "6.33–Do you feel that using beer is harmful to your health?" "6.34–Do you feel that using wine coolers is harmful to your health?" and "6.35–Do you feel that using liquor is harmful to your health?" in *PRIDE Questionnaire Report: 1999–2000 National Summary, Grades 6 Through 12*, PRIDE Surveys, Bowling Green, KY, 2000

FIGURE 6.1

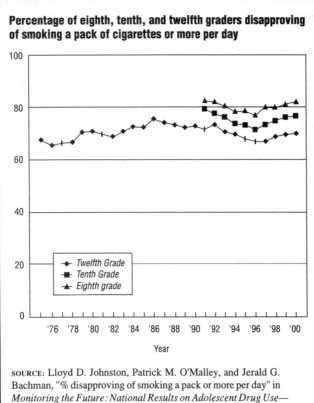

Percentage of eighth, tenth, and twelfth graders disapproving of smoking a pack of cigarettes or more per day

SOURCE: Lloyd D. Johnston, Patrick M. O'Malley, and Jerald G. Bachman, "% disapproving of smoking a pack or more per day" in *Monitoring the Future: National Results on Adolescent Drug Use—Overview of Key Findings, 2000,* Institute for Social Research, University of Michigan, and the National Institute on Drug Abuse, U.S. Department of Health and Human Services, 2001

these substances dangerous, especially at high levels of use (drinking five or more drinks on the weekend or smoking one or more packs of cigarettes a day).

Harmfulness

The surveys found that students are aware of the potential dangers of using alcohol and tobacco, although many consider them less dangerous than occasional or regular use of illicit drugs such as marijuana and cocaine. Among twelfth-graders in the *Monitoring the Future* survey, the proportion who thought daily drinking was very dangerous decreased from 32.7 percent in 1991 to 21.7 percent in 2000, while the proportion who thought smoking was very dangerous increased slightly, from 69.4 percent in 1991 to 73.1 percent in 2000. Only 58.8 percent of eighth-graders in 2000 saw "great risk" associated with being a pack-a-day smoker, but this rate has increased from 51.6 percent in 1991. (See Table 6.1.)

The PRIDE survey found that a greater proportion of sixth-graders than of other middle school students or high school students considered smoking and using alcohol very harmful. Very few students (3.7 percent of those in grades six through twelve) believed that smoking cigarettes caused no harm to health. (See Table 6.2.)

A slightly smaller proportion of seniors in the *Monitoring the Future* study rated the regular use of alcohol as a risk than did younger students. (See Table 6.1.) In fact, most seniors did not consider taking one or two drinks nearly every day (78.3 percent) or binging on five or more drinks once or twice each weekend (57.3 percent) a "great risk." In light of the fact that adult males may improve their health by taking one or two drinks a day, these data are difficult to interpret. They may simply reflect the knowledge of seniors that light-to-moderate drinking in males may be beneficial to health. Binge drinking, however, is not beneficial to health.

Disapproval

Although a higher percentage of twelfth-graders (73.1 percent) than eighth-graders (58.8 percent) thought smoking a pack of cigarettes a day was harmful (Table 6.1), the rate of disapproval of smoking a pack a day was much higher among eighth-graders (81.9 percent) than among twelfth-graders (70.1 percent). (See Figure 6.1 and Table 6.3.)

Among eighth-graders in the *Monitoring the Future* survey, the proportion of those who disapproved of having five or more drinks once or twice each weekend dropped from 85.2 percent in 1991 to 81.2 percent in 2000, with a low disapproval rate of 79.1 percent in 1996. Disapproval of regular use (one or two drinks every day) fell from 82.2 percent to 77.8 percent over the same period, with a low disapproval rate of 74.1 percent in 1996. Among seniors, the proportion who disapproved of taking five or more drinks each weekend decreased from 67.4 percent in 1991 to 65.2 percent in 2000. Additionally, the proportion of seniors who disapproved of regular use dropped significantly, from 76.5 percent in 1991 to 70 percent in 2000. (See Table 6.3.)

ALCOHOL USE

Current Drinkers

According to the *Monitoring the Future* survey, the percentage of high school seniors who had one or more drinks in the 30 days prior to the survey dropped by 4 percent from 1991 to 2000—from 54 percent to 50 percent. (See Table 6.4.) Note that the questionnaire was changed in 1993 to redefine a drink as "more than a few sips."

The proportion of eighth- and tenth-graders who had one or more drinks in the 30 days prior to the survey dropped as well from 1991 to 2000, by 2.7 percent and 1.8 percent, respectively. Even so, alcohol use seems to play a part in the lives of a significant proportion of younger students. In 2000, 22.4 percent of eighth-graders and 41 percent of tenth-graders reported that they used alcohol within the month prior to the survey. (See Table 6.4.)

Heavy Drinking

The *Monitoring the Future* survey found that in 2000 nearly one-third (32.3 percent) of twelfth-graders reported

TABLE 6.3

Trends in disapproval of drug use by eighth, tenth, and twelfth graders, 1991–2000

Percentage who "disapprove or "strongly disapprove"[a]

Do you disapprove of people who . . .[b]	8th Grade											10th Grade											12th Grade										
	1991	1992	1993	1994	1995	1996	1997	1998	1999	2000	'99–'00 change	1991	1992	1993	1994	1995	1996	1997	1998	1999	2000	'99–'00 change	Class of 1991	Class of 1992	Class of 1993	Class of 1994	Class of 1995	Class of 1996	Class of 1997	Class of 1998	Class of 1999	Class of 2000	'99–'00 change
Try one or two drinks of an alcoholic beverage (beer, wine, liquor)	51.7	52.2	50.9	47.8	48.0	45.5	45.7	47.5	48.3	48.7	+0.4	37.6	39.9	38.5	36.5	36.1	34.2	33.7	34.7	35.1	33.4	-1.7	29.8	33.0	30.1	28.4	27.3	26.5	26.1	24.5	24.6	25.2	+0.6
Take one or two drinks nearly every day	82.2	81.0	79.6	76.7	75.9	74.1	76.6	76.9	77.0	77.8	+0.8	81.7	81.7	78.6	75.2	75.4	73.8	75.4	74.6	75.4	73.8	-1.6	76.5	75.9	77.8	73.1	73.3	70.8	70.0	69.4	67.2	70.0	+2.8
Have five or more drinks once or twice each weekend	85.2	83.9	83.3	80.7	80.7	79.1	81.3	81.0	80.3	81.2	+0.9	76.7	77.6	74.7	72.3	72.2	70.7	70.2	70.5	69.9	68.2	-1.7	67.4	70.7	70.1	65.1	66.7	64.7	65.0	63.8	62.7	65.2	+2.5
Smoke one or more packs of cigarettes per day[c]	82.8	82.3	80.6	78.4	78.6	77.3	80.3	80.0	81.4	81.9	+0.5	79.4	77.8	76.5	73.9	73.2	71.6	73.8	75.3	76.1	76.7	+0.6	71.4	73.5	70.6	69.8	68.2	67.2	67.1	68.8	69.5	70.1	+0.6
Use smokeless tobacco regularly	79.1	77.2	77.1	75.1	74.0	74.1	76.5	76.3	78.0	79.2	+1.1	75.4	74.6	73.8	71.2	71.0	71.0	72.3	73.2	75.1	75.8	+0.7	—	—	—	—	—	—	—	—	—	—	—
Approx. N (in thousands) =	17.4	18.5	18.4	17.4	17.6	18.0	18.4	18.1	16.7	16.7		14.8	14.8	15.3	15.9	17.0	15.6	15.7	15.0	13.6	14.3		2.5	2.6	2.7	2.6	2.6	2.4	2.6	2.5	2.3	2.2	

Notes: Level of significance of difference between the two most recent classes: s = .05, ss = .01, sss = .001. '—' indicates data not available. Any apparent inconsistency between the change estimate and the prevalence of use estimates for the two years is due to rounding error.

[a] Answer alternatives were: (1) Don't disapprove, (2) Disapprove, and (3) Strongly disapprove. For 8th and 10th grades, there was another category—"Can't say, drug unfamiliar"—which was included in the calculation of these percentages.
[b] The question asked of twelfth graders was "Do you disapprove of people (who are 18 or older) doing each of the following?"
[c] Beginning in 1999, data based on two-thirds of N indicated due to changes in questionnaire forms.

SOURCE: Lloyd D. Johnston, Patrick M. O'Malley, and Jerald G. Bachman, "Table 6: Trends in Disapproval of Drug Use by Eighth and Tenth Graders, 1991–2000" and "Table 7: Long-Term Trends in Disapproval of Drug Use by Twelfth Graders," in Monitoring the Future: National Results on Adolescent Drug Use—Overview of Key Findings, 2000, Institute for Social Research, University of Michigan, and the National Institute on Drug Abuse, U.S. Department of Health and Human Services, 2001

TABLE 6.4

Trends in annual and 30-day prevalence of use of various drugs for eighth, tenth, and twelfth graders

Annual

	1991	1992	1993	1994	1995	1996	1997	1998	1999	2000	'99–'00 change
Alcohol[a]											
Any use											
8th Grade	54.0	53.7	51.6 / 45.4	46.8	45.3	46.5	45.5	43.7	43.5	43.1	-0.4
10th Grade	72.3	70.2	69.3 / 63.4	63.9	63.5	65.0	65.2	62.7	63.7	65.3	+1.6
12th Grade	77.7	76.8	76.0 / 72.7	73.0	73.7	72.5	74.8	74.3	73.8	73.2	-0.6
Been Drunk[b]											
8th Grade	17.5	18.3	18.2	18.2	18.4	19.8	18.4	17.9	18.5	18.5	0.0
10th Grade	40.1	37.0	37.8	38.0	38.5	40.1	40.7	38.3	40.9	41.6	+0.7
12th Grade	52.7	50.3	49.6	51.7	52.5	51.9	53.2	52.0	53.2	51.8	-1.4
Cigarettes											
Any use											
8th Grade	—	—	—	—	—	—	—	—	—	—	—
10th Grade	—	—	—	—	—	—	—	—	—	—	—
12th Grade	—	—	—	—	—	—	—	—	—	—	—
Smokeless Tobacco[c]											
8th Grade	—	—	—	—	—	—	—	—	—	—	—
10th Grade	—	—	—	—	—	—	—	—	—	—	—
12th Grade	—	—	—	—	—	—	—	—	—	—	—
Steroids[b]											
8th Grade	1.0	1.1	0.9	1.2	1.0	0.9	1.0	1.2	1.7	1.7	0.0
10th Grade	1.1	1.1	1.0	1.1	1.2	1.2	1.2	1.2	1.7	2.2	+0.5s
12th Grade	1.4	1.1	1.2	1.3	1.5	1.4	1.4	1.7	1.8	1.7	-0.1

Approximate Weighted Ns

	1991	1992	1993	1994	1995	1996	1997	1998	1999	2000
8th Grade	17,500	18,600	18,300	17,300	17,500	17,800	18,600	18,100	16,700	16,700
10th Grade	14,800	14,800	15,300	15,800	17,000	15,600	15,500	15,000	13,600	14,300
12th Grade	15,000	15,800	16,300	15,400	15,400	14,300	15,400	15,200	13,600	12,800

30-Day

	1991	1992	1993	1994	1995	1996	1997	1998	1999	2000	'99–'00 change
Alcohol[a]											
Any use											
8th Grade	25.1	26.1	26.2 / 24.3	25.5	24.6	26.2	24.5	23.0	24.0	22.4	-1.7
10th Grade	42.8	39.9	41.5 / 38.2	39.2	38.8	40.4	40.1	38.8	40.0	41.0	+0.9
12th Grade	54.0	51.3	51.0 / 48.6	50.1	51.3	50.8	52.7	52.0	51.0	50.0	-1.0
Been Drunk[b]											
8th Grade	7.6	7.5	7.8	8.7	8.3	9.6	8.2	8.4	9.4	8.3	-1.1
10th Grade	20.5	18.1	19.8	20.3	20.8	21.3	22.4	21.1	22.5	23.5	+1.0
12th Grade	31.6	29.9	28.9	30.8	33.2	31.3	34.2	32.9	32.9	32.3	-0.6
Cigarettes											
Any use											
8th Grade	14.3	15.5	16.7	18.6	19.1	21.0	19.4	19.1	17.5	14.6	-2.8sss
10th Grade	20.8	21.5	24.7	25.4	27.9	30.4	29.8	27.6	25.7	23.9	-1.8
12th Grade	28.3	27.8	29.9	31.2	33.5	34.0	36.5	35.1	34.6	31.4	-3.2ss
Smokeless Tobacco[c]											
8th Grade	6.9	7.0	6.6	7.7	7.1	7.1	5.5	4.8	4.5	4.2	-0.3
10th Grade	10.0	9.6	10.4	10.5	9.7	8.6	8.9	7.5	6.5	6.1	-0.5
12th Grade	—	11.4	10.7	11.1	12.2	9.8	9.7	8.8	8.4	7.6	-0.7
Steroids[b]											
8th Grade	0.4	0.5	0.5	0.5	0.6	0.4	0.5	0.5	0.7	0.8	+0.1
10th Grade	0.6	0.6	0.5	0.6	0.6	0.5	0.7	0.6	0.9	1.0	0.0
12th Grade	0.8	0.6	0.7	0.9	0.7	0.7	1.0	1.1	0.9	0.8	-0.1

Notes: Level of significance of difference between the two most recent classes: s = .05, ss = .01, sss = .001. '—' indicates data not available. Any apparent inconsistency between the change estimate and the prevalence of use estimates for the two most recent classes is due to rounding error.

a For all grades: In 1993, the question text was changed slightly in half of the forms to indicate that a "drink" meant "more than a few sips." The data in the upper line for alcohol came from forms using the original wording, while the data in the lower line came from forms using the revised wording. In 1993, each line of data was based on one of two forms for the 8th and 10th graders and on three of six forms for the 12th graders. N is one-half of N indicated for all groups. Beginning in 1994, data were based on all forms for all grades.

b For 12th graders only: Data based on two of six forms; N is two-sixths of N indicated.

c For 8th and 10th graders only: MDMA data based on one form in 1996; N is one-half of N indicated. Beginning in 1997, data based on one-third of N indicated due to changes on the questionnaire forms. Smokeless tobacco data based on one of two forms for 1991–96 and on two of four forms beginning in 1997; N is one-half of N indicated. For 12th graders only: Data based on one form; N is one-sixth of N indicated.

SOURCE: Lloyd D. Johnston, Patrick M. O'Malley, and Jerald G. Bachman, "Table 2: Trends in Annual and 30-Day Prevalence of Use of Various Drugs for Eighth, Tenth, and Twelfth Graders" and "Footnotes for Table 1 to Table 3" in *Monitoring the Future: National Results on Adolescent Drug Use—Overview of Key Findings, 2000*, Institute for Social Research, University of Michigan, and the National Institute on Drug Abuse, U.S. Department of Health and Human Services, 2001

FIGURE 6.2

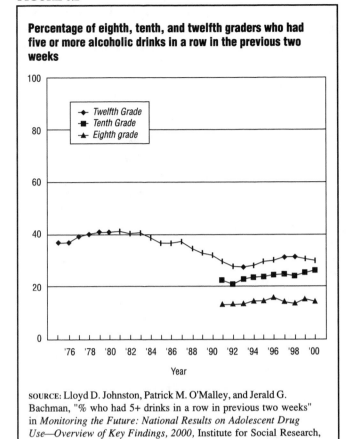

Percentage of eighth, tenth, and twelfth graders who had five or more alcoholic drinks in a row in the previous two weeks

SOURCE: Lloyd D. Johnston, Patrick M. O'Malley, and Jerald G. Bachman, "% who had 5+ drinks in a row in previous two weeks" in *Monitoring the Future: National Results on Adolescent Drug Use—Overview of Key Findings, 2000,* Institute for Social Research, University of Michigan, and the National Institute on Drug Abuse, U.S. Department of Health and Human Services, 2001

FIGURE 6.3

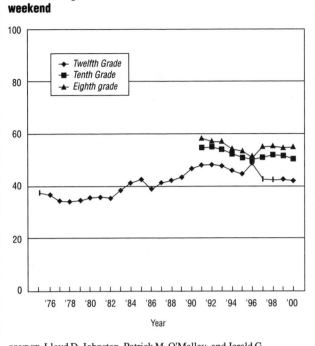

Percent of eigth, tenth, and twelth graders who see "great risk" in having five or more drinks in a row once or twice each weekend

SOURCE: Lloyd D. Johnston, Patrick M. O'Malley, and Jerald G. Bachman, "% seeing "great risk" in having 5+ drinks in a row once or twice each weekend" in *Monitoring the Future: National Results on Adolescent Drug Use—Overview of Key Findings, 2000,* Institute for Social Research, University of Michigan, and the National Institute on Drug Abuse, U.S. Department of Health and Human Services, 2001

being drunk at least once in the month before the survey. Additionally, 23.5 percent of tenth-graders and 8.3 percent of eighth-graders reported being drunk in the past 30 days. (See Table 6.4.)

As Figure 6.2 shows, twelfth-graders reported slight increases in binge or heavy drinking from 1992 to 1998. In 2000 about 30 percent of twelfth-graders reported taking five or more drinks in a row in the two weeks prior to the survey. Binge drinking also occurs among eighth- and tenth-graders, and the proportion of tenth-graders who binge drink increased steadily during the 1990s. Figure 6.3 shows the percentage of eighth-, tenth-, and twelfth-graders who see "great risk" in binge drinking once or twice each weekend. About 56 percent of eighth-graders, 51 percent of tenth-graders, and 43 percent of tenth-graders find this practice risky.

Drinking and Driving

In *Traffic Safety Facts 1999—Young Drivers* (Washington, D.C., 2000) the National Highway Traffic Safety Administration (NHTSA) of the U.S. Department of Transportation reported that the proportion of intoxicated teenage drivers involved in fatal car crashes decreased 41 percent between 1989 and 1999. The NHTSA estimates

that, between 1975 and 1999, the minimum drinking age laws have reduced traffic fatalities involving drivers aged 18–20 by 13 percent, saving approximately 19,121 lives. (See Figure 6.4.)

The NHTSA defines a traffic accident as alcohol-related if either the driver or an involved pedestrian has a blood alcohol concentration (BAC) of 0.01 grams per deciliter (g/dl) of blood or greater. In most states, persons with a BAC of 0.10 g/dl or higher are considered intoxicated. As of September 2001, 23 states plus the District of Columbia have lowered the BAC limit to 0.08.

By the end of 1998, all states and the District of Columbia had enacted "zero-tolerance" laws for drinking drivers under age 21. It is illegal for drivers under 21 to drive with BAC levels of 0.02 g/dl or greater. The National Conference of State Legislatures reported that early evidence on the results of the laws is encouraging. In the first four states to adopt the zero-tolerance standard, nighttime alcohol-related fatal crashes involving drivers under age 21 dropped 34 percent.

When alcohol is involved, the severity of a traffic accident generally increases. In 1999 a total of 8,175 young drivers (ages 15–20) were involved in fatal crashes.

FIGURE 6.4

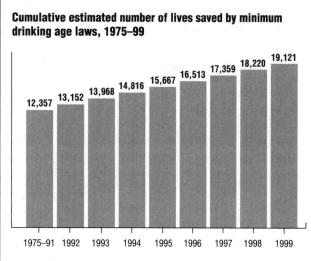

Cumulative estimated number of lives saved by minimum drinking age laws, 1975–99

Year	Value
1975–91	12,357
1992	13,152
1993	13,968
1994	14,816
1995	15,667
1996	16,513
1997	17,359
1998	18,220
1999	19,121

SOURCE: "Figure 3. Cumulative Estimated Number of Lives Saved by Minimum Drinking Age Laws, 1975–1999," in *Traffic Safety Facts 1999—Alcohol*, National Highway Traffic Safety Administration, National Center for Statistics and Analysis, Washington, D.C., 2000

Of those who survived, 7 percent had BAC levels from 0.01 to 0.09 g/dl, and 7 percent had BAC levels of 0.10 g/dl or higher. For those fatally injured, 8 percent had BAC levels between 0.01 and 0.09 g/dl, and 21 percent had BAC levels of 0.10 g/dl or higher.

Alcoholic Novelty Drinks

Many parents are becoming concerned about a new trend in liquor marketing—alcoholic drinks that are particularly attractive to teenagers and young adults. Alcoholic novelties, such as Tooters, are potent, single-serve cocktails. With trendy names like Yellin Melon Balls, Bahama Mama, and Blu-Dacious Kamikaze, and with catchy packaging, the drinks definitely interest young people.

Tooters Lingo Liqueurs are sold in party packs of 30 foil-sealed tubes filled with a single shot of fruit-colored liquid having an alcoholic content of 15 percent. Ray Byrd, operations manager for Tooters, does not believe attractive packaging causes either overconsumption or underage consumption. Tooters "are designed for people who are drinking to have fun, not to get drunk. Our product isn't a slamming type of drink." Critics of alcoholic novelties disagree. Karolyn Nunnallee, president of Mothers Against Drunk Driving (MADD), says, "There aren't a lot of adults that relate to frogs or lizards in advertising, and I don't think mature adults are the ones choosing to use Jell-O shots."

Most alcoholic novelties range in alcohol content from less than 4 percent to 12 percent. Under current law, alcohol can be sold in single-serve containers as long as they meet the Bureau of Alcohol, Tobacco, and Firearms (ATF) regulations. However, a new regulation being con-sidered by the ATF would prohibit companies from packaging alcoholic beverages in containers that could be misleading or that are "likely to be confused with other non-alcohol" products. As of June 2001, the regulation was being revised and was not yet in effect.

TOBACCO USE

Health Consequences of Early Tobacco Use

The likelihood of future health problems due to the use of tobacco, especially cigarettes, is a matter of great concern. According to the Centers for Disease Control and Prevention (CDC) of the U.S. Department of Health and Human Services (HHS), smoking and the use of smokeless tobacco are both closely associated with health problems such as heart disease, lung disease, and cancer.

John K. Wiencke et al., in "Early Age at Smoking Initiation and Tobacco Carcinogen DNA Damage in the Lung" (*Journal of National Cancer Institute,* April 1999), indicates that the age at which smoking is initiated is a significant factor in the risk of lung cancer. Smoking in the teen years appears to cause permanent genetic changes in the lungs, increasing the risk of lung cancer—even if the smoker quits. The younger the smoking starts, the more damage is done. Such damage is less likely among smokers who start in their twenties.

An earlier study, "Preventing Tobacco Use among Young People: A Report of the Surgeon General—Executive Summary" (*Morbidity and Mortality Weekly Report,* March 1994), indicated that cigarette smoking during adolescence seems to retard lung growth and reduce maximum lung function. As a result, youthful smokers are less likely than their nonsmoking peers to be physically fit and more likely to experience shortness of breath, coughing spells, wheezing, and overall poorer health. These health problems pose a clear risk for other chronic conditions in adulthood, such as chronic obstructive pulmonary disease, including emphysema and chronic bronchitis. Early smoking has also been linked to an increased risk of cardiovascular diseases, such as high cholesterol and triglyceride levels, atherosclerosis (arterial plaque), and early onset of heart disease.

Smokeless tobacco also has undesirable health effects on young users. Adolescent use is linked to the development of periodontal disease, soft-tissue damage, and oral cancers. In addition, young people who use smokeless tobacco are more likely than their non-using peers to become cigarette smokers.

Cigarettes

In the *Monitoring the Future 2000* survey, and as shown in Table 6.4, the prevalence of cigarette smoking rose significantly for eighth-, tenth-, and twelfth-graders from 1991 to 1996, and declined from 1997 to 2000. However, the decline still has not reached 1991 levels. For

TABLE 6.5

High school students who used tobacco—from United States Youth Risk Survey, 1999

Category	Lifetime cigarette use[1]			Lifetime daily cigarette use[2]			Current cigarette use[3]			Current frequent cigarette use[4]			Smoked >10 cigarettes/day[5]		
	Female	Male	Total	Female	Male	Total	Female	Male	Total	Female	Male	Total	Female	Male	Total
Race/Ethnicity															
White[6]	70.9 (±3.6)[7]	70.8 (±5.3)	70.9 (±4.1)	29.2 (±3.5)	29.3 (±3.5)	29.3 (±3.2)	39.1 (±3.5)	38.2 (±3.7)	38.6 (±3.2)	19.4 (±3.7)	20.9 (±3.2)	20.2 (±3.0)	4.9 (±1.4)	8.4 (±1.6)	6.6 (±1.2)
Black[6]	68.9 (±6.8)	69.0 (±7.9)	68.9 (±6.7)	8.0 (±3.3)	14.6 (±11.9)	11.2 (±6.9)	17.7 (±3.5)	21.8 (±7.1)	19.7 (±4.1)	5.0 (±3.1)	9.1 (±4.4)	7.0 (±3.4)	1.0 (±0.9)	0.8 (±0.5)	0.9 (±0.6)
Hispanic	71.1 (±4.1)	74.9 (±3.9)	72.9 (±3.2)	18.2 (±3.7)	21.1 (±5.0)	19.6 (±3.2)	31.5 (±4.6)	34.0 (±4.5)	32.7 (±3.8)	8.5 (±3.1)	12.5 (±4.6)	10.4 (±2.7)	2.0 (±1.6)	3.5 (±1.5)	2.7 (±1.1)
Grade															
9	60.3 (±7.2)	63.1 (±5.4)	61.8 (±5.6)	17.3 (±4.4)	19.7 (±4.3)	18.5 (±3.4)	29.2 (±4.8)	26.1 (±6.1)	27.6 (±4.0)	11.0 (±2.8)	11.4 (±3.1)	11.2 (±2.6)	3.0 (±1.2)	3.5 (±1.9)	3.3 (±1.2)
10	75.1 (±2.7)	72.7 (±6.8)	73.9 (±4.1)	27.7 (±3.2)	26.3 (±4.8)	27.0 (±2.9)	35.7 (±4.1)	33.6 (±2.8)	34.7 (±2.5)	15.3 (±4.0)	15.0 (±3.9)	15.2 (±3.5)	3.0 (±1.5)	5.2 (±2.5)	4.1 (±1.2)
11	71.8 (±2.8)	68.1 (±5.2)	69.9 (±3.2)	26.9 (±4.0)	24.4 (±5.3)	25.7 (±3.1)	35.6 (±5.2)	36.4 (±5.9)	36.0 (±3.0)	17.1 (±3.0)	20.4 (±5.7)	18.7 (±2.7)	3.8 (±1.5)	6.1 (±2.7)	4.9 (±1.6)
12	75.5 (±5.4)	80.5 (±3.4)	78.0 (±4.0)	28.8 (±6.0)	34.3 (±8.6)	31.5 (±6.5)	40.5 (±5.9)	45.2 (±6.7)	42.8 (±5.5)	20.3 (±6.0)	26.1 (±10.1)	23.1 (±6.7)	7.2 (±3.5)	10.8 (±5.2)	8.9 (±4.1)
Total	**70.2 (±2.9)**	**70.5 (±3.8)**	**70.4 (±3.0)**	**24.8 (±2.4)**	**25.8 (±3.1)**	**25.3 (±2.6)**	**34.9 (±2.6)**	**34.7 (±3.0)**	**34.8 (±2.5)**	**15.6 (±2.6)**	**17.9 (±3.2)**	**16.8 (±2.5)**	**4.1 (±1.2)**	**6.3 (±1.4)**	**5.2 (±1.2)**

[1]Ever tried cigarette smoking, even one or two puffs.
[2]Ever smoked ≥1 cigarettes every day for 30 days.
[3]Smoked cigarettes on ≥1 of the 30 days preceding the survey.
[4]Smoked cigarettes on ≥20 of the 30 days preceding the survey.
[5]Smoked >10 cigarettes/day on the days smoked during the 30 days preceding the survey.
[6]Non-Hispanic.
[7]Ninety-five percent confidence interval.

SOURCE: "Table 14: Percentage of high school students who used tobacco, by sex, race/ethnicity, and grade–United States, Youth Risk Survey, 1999" from "Youth Risk Behavior Surveillance–United States, 1999," in *Morbidity and Mortality Weekly Report*, June 9, 2000, vol. 49, no. SS-5

eighth-graders, 30-day prevalence rates (the proportion of those who used cigarettes in the 30 days prior to the survey) grew from 14.3 percent in 1991 to 21 percent in 1996, dropping to 14.6 percent in 2000. For tenth-graders, the increase was from 20.8 percent to 30.4 percent during the same period, falling to 23.9 percent in 2000. For twelfth-graders, the rate rose from 28.3 percent in 1991 to a high of 36.5 percent in 1997, declining to 31.4 percent in 2000.

The results of the 1999 *Youth Risk Behavior Surveillance* (*Morbidity and Mortality Weekly Report,* June 9, 2000) showed that 34.8 percent of high school students were current cigarette users. Prevalence of current use was lowest in grade 9, and increased through grades 10, 11, and 12. A greater percentage of females than males reported current cigarette use in grades 9 and 10; a greater percentage of males than females reported current cigarette use in grades 11 and 12. As has been shown in other studies, a greater percentage of white high school students were current smokers than black or Hispanic students. Black students had the lowest prevalence rate. (See Table 6.5.)

PROJECTED SMOKING-RELATED DEATHS. In "Projected Smoking-Related Deaths among Youth—United States" (*Morbidity and Mortality Weekly Report,* November 8, 1996) the CDC projected the future health impact of smoking on youths age 17 and younger. If their present patterns of tobacco use continue, the CDC estimates that more than 5.3 million of these young users will die prematurely in adulthood from smoking-related illnesses. These deaths would account for approximately 64 million years of potential life lost (about 12–21 years per smoker). The overall cost of these early illnesses and deaths is estimated by the CDC at about $200 billion ($1,200 per smoker) in health costs and lost productivity.

Smokeless Tobacco

Smokeless tobacco (primarily snuff and chewing tobacco) is much less widely used than cigarettes. However, because its users often progress to cigarette smoking, medical health practitioners consider smokeless tobacco to be dangerous for young people. In addition, the potential health hazards of early use of chewing tobacco and snuff are well documented. The *Monitoring the Future* survey indicates that current use (within 30 days before the survey) of smokeless tobacco has generally declined since the mid-1990s. (See Table 6.4.)

GENDER, RACIAL, AND ETHNIC DIFFERENCES

Alcohol

In 1999 results from the *Monitoring the Future* survey revealed that alcohol was used by about half (51 percent) of all high school seniors. About 55 percent of male seniors and 47 percent of female seniors reported using alcohol one or more times in the 30 days prior to the survey. About 38 percent of male seniors and 28 percent of female seniors said they had been drunk at least once during the 30-day period before the survey. Rates of alcohol use and drunkenness dropped as grade level dropped. (See Table 6.6.)

The rate of alcohol use in the 30 days prior to the survey was higher among white high school seniors (56.3 percent) than among black (32.2 percent) and Hispanic seniors (50.2 percent). However, among eighth-graders, Hispanic students were the most likely to report alcohol use—29 percent of Hispanics, compared with 24.7 percent of white students and 16.1 percent of black students, reported alcohol use in the last 30 days. (See Table 6.7.)

Cigarettes

More than one-third (34.6 percent) of all high school seniors reported that they had smoked one or more cigarettes in the 30 days before the survey, and 8.4 percent reported using smokeless tobacco. The rate at which male seniors reported smoking was only slightly higher than the female senior rate (35.4 percent versus 33.5 percent). In the lower grades, the ratio of male-to-female smoking was about equal as well. (See Table 6.6.)

Male seniors (15.5 percent) used smokeless tobacco at about 12 times the rate of female seniors (1.3 percent). While the gender difference was less pronounced in the eighth grade, the rate of female initiation or continuation of the use of smokeless tobacco dropped as the girls grew older. The male rate increased as the boys grew older. (See Table 6.6.)

White seniors smoked at a higher rate in the 30 days prior to the survey (40.1 percent) than did Hispanic (27.3 percent) or black (14.9 percent) seniors. On the other hand, Hispanic eighth-graders (20.5 percent) smoked at about the same rate as white eighth-graders (20.1 percent). White seniors were more likely than blacks or Hispanics to use cigarettes daily. (See Table 6.7.)

White high school seniors (11 percent) were about three times as likely as Hispanic seniors (3.9 percent), and about seven times as likely as black seniors (1.5 percent), to have used smokeless tobacco within 30 days prior to the survey. Black students were the least likely to use chewing tobacco and snuff at all grade levels. (See Table 6.7.)

AVAILABILITY OF ALCOHOL AND TOBACCO TO MINORS

The sale of alcohol to persons under the age of 21 and tobacco to persons under the age of 18 is illegal. However, results of surveys cited in this chapter show that an overwhelming majority of young persons reported that both were readily available to them.

According to the 1999–2000 PRIDE survey, 42.2 percent of junior high students and 79.5 percent of high school students found it fairly easy or very easy to get cigarettes. Among alcoholic beverages, beer was most available, with 39.6 percent of junior high students and 73.5

TABLE 6.6

Thirty-day prevalence of use of alcohol, cigarettes, and smokeless tobacco by eighth, tenth, and twelfth graders, 1999

(Entries are percentages)

	Alcohol			Been Drunk[1]			Cigarettes			Smokeless Tobacco[2]		
	8th	10th	12th	8th	10th	12th	8th	10th	12th	8th	10th	12th
Total	24.0	40.0	51.0	9.4	22.5	32.9	17.5	25.7	34.6	4.6	6.5	8.4
Sex:												
Male	24.8	42.3	55.3	10.2	25.4	37.9	16.7	25.2	35.4	6.9	12.2	15.5
Female	23.3	38.1	46.8	8.6	19.8	27.7	17.7	25.8	33.5	2.1	1.3	1.3
College Plans:												
None or under 4 yrs.	41.6	53.7	55.2	22.4	34.6	36.1	40.3	44.0	44.9	13.2	13.2	10.5
Complete 4 yrs.	22.0	37.9	49.8	8.0	20.7	31.7	14.5	22.7	31.4	3.5	5.4	7.6
Region:												
Northeast	25.7	44.8	57.2	9.4	25.8	37.5	15.7	28.0	34.2	2.5	5.2	4.3
North Central	25.7	40.9	51.1	11.6	26.0	33.4	21.3	30.2	37.8	5.3	8.1	8.9
South	24.4	38.8	49.5	9.5	20.3	30.8	18.7	26.3	36.2	5.9	7.9	10.7
West	19.8	36.1	47.8	6.6	19.0	32.2	12.1	17.5	27.6	2.9	4.0	7.0
Population Density:												
Large MSA	21.7	39.7	48.9	7.8	21.6	29.2	12.7	22.9	30.0	1.8	4.6	4.9
Other MSA	23.4	39.7	52.8	8.4	22.7	35.4	16.0	25.0	35.0	3.9	5.3	8.5
Non-MSA	28.1	41.0	50.1	13.3	23.4	32.5	26.1	30.4	38.7	8.9	11.3	11.7
Parental Education:[3]												
1.0-2.0 (Low)	30.7	40.6	46.8	14.5	21.8	20.8	26.6	30.5	33.0	6.6	7.2	5.4
2.5-3.0	27.9	42.3	50.5	11.7	23.4	30.5	23.5	29.6	37.3	5.7	7.0	9.1
3.5-4.0	25.2	40.2	51.1	9.9	23.3	34.0	17.0	26.0	35.0	4.5	7.3	8.8
4.5-5.0	20.4	38.7	50.2	6.9	21.7	32.8	12.3	22.4	32.4	3.3	6.1	8.5
5.5-6.0 (High)	22.1	40.9	56.0	8.7	24.0	40.6	12.2	21.4	34.4	3.1	4.8	7.9

NOTE: '—' indicates data not available.

[1] 12th grade only: Data based on two of six forms; N is two-sixths of N indicated.

[2] 8th and 10th grade only: Data based on two of four forms; N is one-half of N indicated. 12th grade only: Data based on one of six forms; N is one-sixth of N indicated.

[3] Parental education is an average score of mother's education and father's education reported on the following scale: (1) Completed grade school or less, (2) Some high school, (3) Completed high school, (4) Some college, (5) Completed college, (6) Graduate or professional school after college. Missing data was allowed on one of the two variables.

SOURCE: Lloyd D. Johnston, Patrick M. O'Malley, and Jerald G. Bachman, "Table 4-7: Thirty-Day Prevalence of Use of Various Drugs by Subgroups Eighth, Tenth, and Twelfth Graders, 1999," in *Monitoring the Future: National Survey Results on Drug Use, 1975–1999: Volume 1: Secondary School Students 1999* Institute for Social Research, University of Michigan, and the National Institute on Drug Abuse, U.S. Department of Health and Human Services, 2000

percent of high school students saying that it was fairly or very easy to obtain. A significant proportion of junior high students found wine coolers (36.1 percent) and liquor (29 percent) fairly or very easy to get, as did about 7 out of 10 high school students (71.5 percent and 67.5 percent, respectively). (See Table 6.8.)

Results from the 1999 *Youth Risk Behavior Surveillance* revealed that over two-thirds (69.6 percent) of student smokers under age 18 were able to purchase cigarettes without being asked to show proof of age. About one-fourth (23.5 percent) of underage smokers had purchased cigarettes at a store or gas station in the 30 days preceding the survey. (See Table 6.9.)

COLLEGE STUDENTS AND OTHER YOUNG ADULTS

The *Monitoring the Future* survey also questions college students and other young adults (ages 19–28) not in college about various risk behaviors. Results of the 1999 survey show that the daily-use rates for alcohol were similar between college students and young adults. Both groups showed a decrease in prevalence rates for alcohol

use from 1991 to 1995, but that trend changed from 1996 to 1999. By 1999, the prevalence of daily alcohol use increased to 4.5 percent for college students and to 4.8 percent for young adults not in college. (See Table 6.10.)

A greater percentage of college students, however, engaged in binge drinking (drinking five or more drinks in a row at one time) within the two-week period preceding the survey than did their noncollege peers. In 1999, 40 percent of college students reported binge drinking within the two weeks prior to the survey, compared with 35.8 percent of those not attending college. From 1998 to 1999, the percentage of students engaging in binge drinking increased slightly for both groups. (See Table 6.10.)

Cigarette use among college students and their noncollege peers has increased since 1997. In 1999 the percentage of young adults who smoked cigarettes daily was 21.5 percent, up from 20.6 in 1997. The percentage of college students who smoked every day was 19.3 in 1999, compared with 15.2 in 1997. A higher percentage of young persons not in college smoked a half-pack or more per day (15.1 percent) than did college students (11 percent). (See Table 6.10.)

TABLE 6.7

Racial/ethnic comparisons of lifetime, annual, thirty-day, and daily prevalence of use of alcohol, cigarettes, and smokeless tobacco among eighth, tenth, and twelfth graders

NOTE: Percentages are based on 1998 and 1999 data combined.[1]

	Alcohol			Been Drunk[2]			5+ Drinks[3]			Cigarettes			Smokeless Tobacco[4]		
	8th	10th	12th	8th	10th	12th	8th	10th	12th	8th	10th	12th	8th	10th	12th
Lifetime:															
White	52.3	72.3	82.9	26.0	52.2	67.2	—	—	—	45.5	60.8	68.9	17.3	25.8	31.4
Black	48.1	60.7	71.8	17.8	29.5	40.6	—	—	—	41.0	43.9	47.2	7.4	7.6	5.1
Hispanic	59.6	73.2	82.8	29.3	47.4	62.7	—	—	—	49.4	58.2	63.9	12.7	15.8	13.4
Annual:															
White	45.1	66.5	77.5	19.9	44.8	58.4	—	—	—	—	—	—	—	—	—
Black	33.7	48.8	60.0	10.3	19.9	29.0	—	—	—	—	—	—	—	—	—
Hispanic	50.6	65.8	74.7	21.0	36.5	49.1	—	—	—	—	—	—	—	—	—
30-Day:															
White	24.7	43.0	56.3	9.8	25.7	37.8	—	—	—	20.1	30.8	40.1	5.4	8.7	11.0
Black	16.1	24.4	32.2	4.9	7.6	14.9	—	—	—	10.7	12.5	14.9	2.3	1.6	1.5
Hispanic	29.0	39.6	50.2	9.9	17.8	27.5	—	—	—	20.5	21.1	27.3	4.6	4.8	3.9
Daily:															
White	0.8	1.9	3.9	—	—	—	14.3	27.2	35.7	9.7	19.1	26.9	1.1	2.4	4.3
Black	0.9	1.0	1.6	—	—	—	9.9	12.7	12.3	3.8	5.3	7.7	0.4	0.3	0.0
Hispanic	1.3	2.4	4.8	—	—	—	20.9	27.5	29.3	8.5	9.1	14.0	1.0	0.8	0.4

NOTE: '—' indicates data not available.

[1] To derive percentages for each racial subgroup, data for the specified year and the previous year have been combined to increase subgroup sample sizes and thus provide more stable estimates.

[2] 12th grade only: Data based on one of six forms; N is two-sixth of N indicated.

[3] This measure refers to having five or more drinks in a row in the last two weeks.

[4] 8th and 10th grade only: Data based on two of four forms; N is one-half of N indicated. 12th grade only: Data based on one of six forms; N is one-sixth of N indicated.

SOURCE: Lloyd D. Johnston, Patrick M. O'Malley, and Jerald G. Bachman, "Table 4-9: Racial/Ethnic Comparisons of Lifetime, Annual, Thirty-Day, and Daily Prevalence of Use of Various Drugs: Eighth, Tenth, and Twelfth Graders," in *Monitoring the Future: National Survey Results on Drug Use*, 1975–1999: Volume 1: Secondary School Students 1999 *Institute for Social Research*, University of Michigan, and the National Institute on Drug Abuse, U.S. Department of Health and Human Services, 2000

TABLE 6.8

PRIDE Survey: Students' responses about ease of obtaining tobacco and alcohol

How easy is it to get cigarettes?

Grade Level	N of Valid	N of Miss	Cannot Get	Very Difficult	Fairly Difficult	Fairly Easy	Very Easy
6th	17512	1190	60.6	6.1	5.7	11.3	16.4
7th	16000	1056	47.0	5.1	7.7	16.4	23.8
8th	22174	1311	33.4	4.0	7.5	20.8	34.4
9th	15111	844	22.7	2.3	5.6	21.8	47.7
10th	15789	967	16.7	1.4	3.7	22.2	56.0
11th	10138	546	12.3	1.1	2.9	19.6	64.3
12th	11207	473	7.6	0.6	1.3	12.9	77.6
JrHs	55686	3557	45.8	5.0	7.0	16.5	25.7
SrHS	52245	2830	15.6	1.4	3.6	19.6	59.9
Total	107931	6387	31.2	3.2	5.3	18.0	42.2

How easy is it to get beer?

Grade Level	N of Valid	N of Miss	Cannot Get	Very Difficult	Fairly Difficult	Fairly Easy	Very Easy
6th	17364	1338	60.5	6.9	6.3	10.3	16.1
7th	15863	1193	48.8	6.2	8.0	13.9	23.0
8th	22061	1424	33.8	5.3	8.8	18.8	33.3
9th	15016	939	23.6	3.6	8.6	22.2	41.9
10th	15687	1069	16.5	2.6	7.6	25.7	47.5
11th	10087	597	13.5	2.1	6.7	26.6	51.1
12th	11142	538	9.8	1.7	6.0	28.5	54.0
JrHs	55288	3955	46.5	6.1	7.8	14.7	24.9
SrHS	51932	3143	16.5	2.6	7.4	25.5	48.0
Total	107220	7098	32.0	4.4	7.6	19.9	36.1

How easy is it to get wine coolers?

Grade Level	N of Valid	N of Miss	Cannot Get	Very Difficult	Fairly Difficult	Fairly Easy	Very Easy
6th	17346	1356	64.2	6.7	6.3	9.6	13.2
7th	15812	1244	51.7	6.5	8.2	13.4	20.2
8th	21998	1487	36.4	5.7	9.5	18.5	29.9
9th	14980	975	25.2	4.0	9.4	22.3	39.1
10th	15670	1086	17.9	2.9	8.5	25.8	45.0
11th	10060	624	14.3	2.4	6.9	26.5	49.8
12th	11148	532	10.2	1.7	6.3	28.7	53.1
JrHs	55156	4087	49.5	6.2	8.1	14.2	21.9
SrHS	51858	3217	17.7	2.9	8.0	25.5	46.0
Total	107014	7304	34.1	4.6	8.1	19.7	33.5

How easy is it to get liquor?

Grade Level	N of Valid	N of Miss	Cannot Get	Very Difficult	Fairly Difficult	Fairly Easy	Very Easy
6th	17312	1390	71.7	6.5	5.5	6.6	9.7
7th	15774	1282	59.1	7.0	7.7	10.1	16.1
8th	21963	1522	43.0	6.6	9.3	15.3	25.8
9th	14945	1010	28.5	5.1	10.3	19.9	36.2
10th	15636	1120	19.5	3.9	9.7	24.7	42.2
11th	10068	616	15.5	3.1	8.3	26.2	46.9
12th	11117	563	10.8	2.6	7.9	28.4	50.3
JrHs	55049	4194	56.6	6.7	7.6	11.1	17.9
SrHS	51766	3309	19.4	3.8	9.2	24.4	43.1
Total	106815	7503	38.6	5.3	8.4	17.5	30.1

SOURCE: "6.131–How easy is it to get cigarettes?" "6.134–How easy is it to get beer?" "6.135–How easy is it to get wine coolers?" and "6.136–How easy is it to get liquor?" in *PRIDE Questionnaire Report: 1999–2000 National Summary, Grades 6 Through 12,* PRIDE Surveys, Bowling Green, KY, 2000

TABLE 6.9

Percentage of high school students under 18 who smoke cigarettes, who obtain cigarettes at a store or gas station, and who purchased cigarettes without being asked to show proof of age, 1999

Category	Purchased cigarettes at a store or gas station[†]				Were not asked to show proof of age when purchasing cigarettes[§]		
	Female	Male	Total		Female	Male	Total
Race/Ethnicity							
White[¶]	17.6	31.5	**24.4**		69.6	63.4	**65.7**
	(±4.9)**	(±5.0)	**(±4.4)**		(±10.9)	(±6.5)	**(±7.5)**
Black[¶]	29.7	31.6	**30.7**		NA[††]	80.2	**84.1**
	(±18.6)	(±10.3)	**(±13.6)**		NA	(±14.2)	**(±13.0)**
Hispanic	15.5	25.0	**20.1**		NA	46.5	**60.7**
	(±8.0)	(±9.1)	**(±6.6)**		NA	(±13.0)	**(±12.6)**
Grade							
9	8.8	15.8	**12.0**		NA	NA	**79.1**
	(±3.7)	(±8.4)	**(±4.8)**		NA	NA	**(±9.9)**
10	16.2	27.9	**21.9**		NA	61.6	**67.9**
	(±8.7)	(±8.1)	**(±7.5)**		NA	(±17.1)	**(±9.9)**
11	21.4	36.1	**28.5**		70.4	67.1	**68.4**
	(±5.6)	(±10.1)	**(±6.4)**		(±15.7)	(±16.9)	**(±13.1)**
12	32.6	44.9	**38.7**		NA	NA	**65.0**
	(±12.0)	(±13.5)	**(±7.0)**		NA	NA	**(±10.9)**
Total	**17.6**	**29.7**	23.5		**76.2**	**65.5**	69.6
	(±5.6)	**(±4.7)**	**(±4.5)**		**(±7.9)**	**(±5.7)**	**(±5.7)**

[†] Purchased cigarettes at a store or gas station during the 30 days preceding the survey.
[§] Among those who purchased cigarettes at a store or gas station during the 30 days preceding the survey.
[¶] Non-Hispanic.
** Ninety-five percent confidence interval.
[††] Not available.

SOURCE: "Table 18: Percentage of high school students aged <18 years who were current cigarette smokers and usually obtained their own cigarettes by purchasing them in a store or gas station and who purchased cigarettes without being asked to show proof of age, by sex, race/ethnicity, and grade," in "Youth Risk Behavior Surveillance—United States, 1999," *Morbidity and Mortality Weekly Report,* vol. 49, no. SS-5, June 9, 2000

TABLE 6.10

Trends in 30-day prevalence of daily use of alcohol and cigarettes for eighth, tenth, and twelfth graders, college students, and young adults (ages 19–28)

	1991	1992	1993	1994	1995	1996	1997	1998	1999	'98-'99 change
Alcohol[1,2]										
Any daily use										
8th grade	0.5	0.6	0.8	—	—	—	—	—	—	—
			1.0	1.0	0.7	1.0	0.8	0.9	1.0	+0.1
10th grade	1.3	1.2	1.6	—	—	—	—	—	—	—
			1.8	1.7	1.7	1.6	1.7	1.9	1.9	0.0
12th grade	3.6	3.4	2.5	—	—	—	—	—	—	—
			3.4	2.9	3.5	3.7	3.9	3.9	3.4	-0.6s
College Students	4.1	3.7	3.9	3.7	3.0	3.2	4.5	3.9	4.5	+0.6
Young Adults	4.9	4.5	4.5	3.9	3.9	4.0	4.6	4.0	4.8	-0.9s
Been Drunk, daily [2,3]										
8th grade	0.1	0.1	0.2	0.3	0.2	0.2	0.2	0.3	0.4	+0.1
10th grade	0.2	0.3	0.4	0.4	0.6	0.4	0.6	0.6	0.7	+0.1
12th grade	0.9	0.8	0.9	1.2	1.3	1.6	2.0	1.5	1.9	+0.4
College Students	—	—	—	—	—	—	—	—	—	—
Young Adults	—	—	—	—	—	—	—	—	—	—
5+ drinks in a row in last 2 weeks										
8th grade	12.9	13.4	13.5	14.5	14.5	15.6	14.5	13.7	15.2	+1.5s
10th grade	22.9	21.1	23.0	23.6	24.0	24.8	25.1	24.3	25.6	+1.3
12th grade	29.8	27.9	27.5	28.2	29.8	30.2	31.3	31.5	30.8	-0.7
College Students	42.8	41.4	40.2	40.2	38.6	38.3	40.7	38.9	40.0	+1.1
Young Adults	34.7	34.2	34.4	33.7	32.6	33.6	34.4	34.1	35.8	+1.7
Cigarettes										
Any daily use										
8th grade	7.2	7.0	8.3	8.8	9.3	10.4	9.0	8.8	8.1	-0.7
10th grade	12.6	12.3	14.2	14.6	16.3	18.3	18.0	15.8	15.9	+0.1
12th grade	18.5	17.2	19.0	19.4	21.6	22.2	24.6	22.4	23.1	+0.7
College Students	13.8	14.1	15.2	18.2	15.8	15.9	15.2	18.0	19.3	+1.3
Young Adults	21.7	20.9	20.8	20.7	21.2	21.8	20.6	21.9	21.5	-0.3
1/2 pack +/day										
8th grade	3.1	2.9	3.5	3.6	3.4	4.3	3.5	3.6	3.3	-0.3
10th grade	6.5	6.0	7.0	7.6	8.3	9.4	8.6	7.9	7.6	-0.3
12th grade	10.7	10.0	10.9	11.2	12.4	13.0	14.3	12.6	13.2	+0.6
College Students	8.0	8.9	8.9	8.0	10.2	8.4	9.1	11.3	11.0	-0.3
Young Adults	16.0	15.7	15.5	15.3	15.7	15.3	14.6	15.6	15.1	-0.5

NOTES: Level of significance of difference between the two years: s=.05, ss=.01, sss=.001.
'—' indicates data not available. '*' indicates less than .05 percent but greater than 0 percent.
Any apparent inconsistency between the change estimate and the prevalence of use estimates for the two years is due to rounding error.

Approximate Weighted Ns	1991	1992	1993	1994	1995	1996	1997	1998	1999
8th Graders	17,500	18,600	18,300	17,300	17,500	17,800	18,600	18,100	16,700
10th Graders	14,800	14,800	15,300	15,800	17,000	15,600	15,500	15,000	13,600
12th Graders	15,000	15,800	16,300	15,400	15,400	14,300	15,400	15,200	13,600
College Students	1,410	1,490	1,490	1,410	1,450	1,450	1,480	1,440	1,440
Young Adults	6,600	6,800	6,700	6,500	6,400	6,300	6,400	6,200	6,000

[1]For 8th, 10th, and 12th graders only: In 1993, the question text was changed slightly in half of the forms to indicate that a "drink" meant "more than just a few sips." The data in the upper line for alcohol came from forms using the original wording, while the data in the lower line came from forms using the revised wording. In 1993, each line of data was based on one of two forms for the 8th and 10th graders and on three of six forms for the 12th graders. N is one-half of N indicated for these groups. Beginning in 1994, data were based on all forms for all grades. For college students and young adults, the revision of the question text resulted in rather little change in the reported prevalence of use. The data for all forms are used to provide the most reliable estimate of change.

[2] Daily used is defined as use on twenty or more occasions in the past thirty days except for cigarettes and smokeless tobacco, for which actual daily use is measured, and for 5+ drinks, for which the prevalence of having five or more drinks in a row in the last two weeks is measured.

[3] For 12th graders, college students, and young adults only: Data based on two of six forms; N is two-sixths of N indicated for each group.

SOURCE: Lloyd D. Johnston, Patrick M. O'Malley, and Jerald G. Bachman, "Table 2-3: Trends in 30-Day Prevalence of Daily Use of Various Drugs for Eighth, Tenth, and Twelfth Graders, College Students, and Young Adults (Ages 19–28)," in *Monitoring the Future: National Survey Results on Drug Use, 1975–1999: Volume 1: Secondary School Students 1999* Institute for Social Research, University of Michigan, and the National Institute on Drug Abuse, U.S. Department of Health and Human Services, 2000

CHAPTER 7
ECONOMICS OF ALCOHOL AND TOBACCO

The alcohol and tobacco industries play large roles in the American economy. The Beer Institute (Washington, D.C.) claims that the brewing industry in 1997–98 provided jobs for about 850,000 brewers, wholesalers, and retailers. Both industries not only provide jobs and income for those involved in growing, manufacturing, and selling these products, but also contribute significant tax revenues to the federal, state, and local governments.

U.S. ALCOHOL SALES AND CONSUMPTION

According to the Economic Research Service of the United States Department of Agriculture, retail sales of alcoholic beverages totaled approximately $104.4 billion in 2000. In 1996 beer sales accounted for 60 percent of the total retail sales, a percentage that has increased steadily since 1972. Distilled spirits accounted for 28.2 percent, falling from about half of all 1967 alcoholic beverage sales. Wine sales have remained fairly stable, ranging between 10 and 13 percent of retail sales over the last 20 years.

Distilled Spirits Sales and Consumption

Over the 12 months ending June 30, 1998, the Distilled Spirits Council reported that 330 million gallons of distilled spirits (hard liquor) were consumed in the United States. In 1997 per capita consumption was 1.24 gallons. According to *Adam's Business Media Liquor Handbook*, liquor consumption dropped 20 percent from 1985 to 1995. Over the next two years, consumption remained fairly constant, increasing by 1 percent. In 1996 U.S. liquor sales totaled $25.6 billion.

Beer Sales and Consumption

The Beer Institute reported that in 1999, an estimated 196 million barrels of domestic and imported beer were shipped to wholesalers in the United States. This represents an increase in shipments of more than 3 million barrels, or 1.6 percent over the 1998 total, and set an all-time

record for beer industry shipments. The 1999 per capita consumption of beer was estimated at 22.3 gallons.

The number of domestic brewers in the United States in September 1997 was 1,698, nearly 12 times the number of brewers in business in 1987. According to the Beer Institute, microbreweries and brewpubs account for this increase. In 1997 Anheuser-Busch, Inc., was again the domestic sales leader, with 46 percent of sales, followed by Miller Brewing Company (22 percent), Stroh Brewery Company (12 percent), Adolph Coors Company (11 percent), and Pabst Brewing Company (3 percent).

Wine Sales and Consumption

In 2000 the United States was the fourth leading wine producer in the world, topped only by Italy, France, and Spain. According to the Wine Institute (San Francisco, California), 565 million gallons of wine were sold in the United States in 2000 for a retail sales value of almost $19 billion. This represents a 3 percent increase over the volume of sales in 1999. The California wine industry accounts for about 90 percent of U.S. production.

While the United States is a leading wine producer, its per capita consumption of wine ranks among the lowest of major wine-producing countries. In 1997 the per capita consumption in Italy and France was 15.3 and 15.9 gallons, respectively, while the U.S. per capita consumption was slightly less than 2 gallons. In 1999 the U.S. per capita consumption was slightly greater than 2 gallons.

HOW MUCH DO INDIVIDUALS AND FAMILIES SPEND ON ALCOHOL?

The U.S. Bureau of Labor Statistics (Washington, D.C., 2000) reported that the average American family spent $318 on alcoholic beverages in 1999, or about 0.85 percent of its average annual expenditures. (See Table 7.1.) This percentage has remained relatively stable since 1993.

TABLE 7.1

Shares of average annual expenditures, Consumer Expenditure Survey, 1999

| Item | Total husband and wife consumer units | Husband and wife only | Husband and wife with children | | | Other husband and wife consumer units | One parent, at least one child under 18 | Single person and other consumer units |
			Total husband and wife with children	Oldest child under 6	Oldest child 6 to 17	Oldest child 18 or over			
Number of consumer units (in thousands)	56,427	23,404	28,535	5,304	15,378	7,853	4,488	6,571	45,467
Income before taxes [1]	$59,128	$54,067	$63,666	$57,922	$63,558	$68,094	$56,519	$25,685	$28,280
Age of reference person	48.2	56.8	41.2	31.7	39.5	50.9	48.4	36.3	49.2
Average number in consumer unit:									
Persons	3.2	2.0	3.9	3.5	4.1	3.9	4.9	2.9	1.6
Earners	1.7	1.2	2.0	1.6	1.8	2.6	2.4	1.1	.9
Vehicles	2.6	2.4	2.7	2.1	2.6	3.2	2.9	1.2	1.3
Percent homeowner	81	84	79	67	79	86	76	39	49
Average annual expenditures	$47,188	$42,185	$51,186	$46,091	$51,493	$54,248	$47,948	$27,918	$25,862
Food	13.5	12.8	13.7	11.7	14.5	13.7	15.5	16.2	13.6
Food at home	7.8	7.1	8.1	7.3	8.5	7.9	9.4	10.5	7.7
Cereals and bakery products	1.2	1.1	1.3	1.1	1.4	1.2	1.4	1.8	1.2
Meats, poultry, fish, and eggs	2.0	1.8	2.0	1.7	2.1	2.1	2.6	2.8	2.0
Dairy products	.9	.8	.9	.9	1.0	.9	1.0	1.1	.8
Fruits and vegetables	1.3	1.3	1.3	1.2	1.4	1.3	1.7	1.6	1.3
Other food at home	2.4	2.1	2.5	2.4	2.6	2.4	2.8	3.2	2.4
Food away from home	5.7	5.6	5.6	4.4	6.0	5.8	6.0	5.7	5.9
Alcoholic beverages	.7	.9	.6	.6	.6	.5	.6	.5	1.2
Housing	31.3	30.7	31.9	37.3	31.9	29.0	30.0	36.2	34.7
Shelter	17.8	17.2	18.4	21.9	18.4	16.3	16.6	21.0	21.1
Utilities, fuels, and public services	6.1	6.1	5.9	5.6	5.8	6.1	6.9	7.9	7.0
Household operations	1.8	1.3	2.2	4.6	2.1	1.0	1.5	2.5	1.7
Housekeeping supplies	1.4	1.5	1.4	1.2	1.5	1.5	1.3	1.3	1.2
Household furnishings and equipment	4.2	4.6	4.0	4.0	4.1	4.1	3.6	3.5	3.8
Apparel and services	4.6	4.0	4.9	4.5	5.2	4.6	5.3	7.0	4.6
Transportation	19.7	19.1	20.0	20.3	18.6	22.2	20.4	16.8	17.5
Vehicle purchases (net outlay)	9.4	9.0	9.7	10.5	9.2	10.0	9.1	8.1	8.0
Gasoline and motor oil	2.9	2.7	3.0	2.8	2.8	3.4	3.4	2.6	2.7
Other vehicle expenses	6.3	6.0	6.4	6.3	5.8	7.7	6.7	5.3	5.7
Public transportation	1.1	1.4	.9	.7	.8	1.1	1.1	.8	1.1
Health care	5.3	6.9	4.3	3.7	4.2	4.8	5.3	3.6	5.4
Entertainment	5.3	5.4	5.4	4.6	6.1	4.7	4.4	4.9	4.6
Personal care products and services	1.1	1.1	1.1	1.0	1.1	1.1	1.0	1.3	1.2
Reading	.4	.5	.4	.3	.4	.4	.3	.3	.5
Education	1.8	1.3	2.2	.7	2.0	3.4	1.2	1.5	1.6
Tobacco products and smoking supplies	.7	.6	.7	.5	.7	.7	1.0	.9	1.1
Miscellaneous	2.2	2.2	2.1	2.4	1.9	2.3	2.8	3.0	2.7
Cash contributions	3.1	4.3	2.3	1.8	2.3	2.7	3.0	1.3	3.7
Personal insurance and pensions	10.2	10.2	10.4	10.7	10.5	9.9	9.2	6.5	7.6
Life and other personal insurance	1.3	1.3	1.3	.9	1.4	1.3	1.3	.6	.7
Pensions and Social Security	8.9	8.9	9.1	9.9	9.1	8.6	7.9	5.9	6.9

[1]Components of income and taxes are derived from "complete income reporters" only.

SOURCE: "Table 49. Composition of consumer unit: Shares of average annual expenditures and sources of income, Consumer Expenditure Survey, 1999," Bureau of Labor Statistics, Washington, DC, 2000

The amount spent on alcohol varied in 1999, depending on the characteristics of the household. On average, single parents with at least one child under age 18 spent only $144 on alcoholic beverages in 1999, or about 0.5 percent of their average annual expenditures. Married couples with no children spent the largest amount, $388, or about 0.9 percent of their average annual expenditures. (See Table 7.2.) The percentage other groups spent on alcohol varied within the 0.5 to 1 percent range, except for the "single person and other consumer units" group. This group spent the most on alcohol: 1.2 percent of their average annual expenditures.

U.S. TOBACCO PRODUCTION AND CONSUMPTION

Farming Trends

Sparked by the invention of the cigarette-making machine, which made cigarettes more affordable, tobacco production in the United States grew from about 300 million pounds in the mid-1860s to over a billion pounds by 1909. By the mid-1940s, tobacco production topped 2 billion pounds as cigarette consumption continued to grow.

During the 1960s, changes in tobacco preparation and the introduction of machinery increased the

TABLE 7.2

Average annual expenditures, Consumer Expenditure Survey, 1999

Item	Total husband and wife consumer units	Husband and wife only	Total husband and wife with children	Oldest child under 6	Oldest child 6 to 17	Oldest child 18 or over	Other husband and wife consumer units	One parent, at least one child under 18	Single person and other consumer units
				Husband and wife with children					
Number of consumer units (in thousands)	56,427	23,404	28,535	5,304	15,378	7,853	4,488	6,571	45,467
Consumer unit characteristics:									
Income before taxes[1]	$59,128	$54,067	$63,666	$57,922	$63,558	$68,094	$56,519	$25,685	$28,280
Age of reference person	48.2	56.8	41.2	31.7	39.5	50.9	48.4	36.3	49.2
Average number in consumer unit:									
Persons	3.2	2.0	3.9	3.5	4.1	3.9	4.9	2.9	1.6
Children under 18	.9	n.a.	1.6	1.5	2.1	.6	1.5	1.8	.2
Persons 65 and over	.3	.7	.1	([2])	([2])	.2	.4	([2])	.3
Earners	1.7	1.2	2.0	1.6	1.8	2.6	2.4	1.1	.9
Vehicles	2.6	2.4	2.7	2.1	2.6	3.2	2.9	1.2	1.3
Percent homeowners	81	84	79	67	79	86	76	39	49
Average annual expenditures	$47,188	$42,185	$51,186	$46,091	$51,493	$54,248	$47,948	$27,918	$25,862
Food	6,372	5,380	7,034	5,379	7,472	7,415	7,419	4,526	3,507
Food at home	3,695	3,000	4,146	3,360	4,381	4,287	4,528	2,942	1,987
Cereals and bakery products	567	446	653	503	713	646	656	509	299
Meats, poultry, fish, and eggs	939	771	1,031	769	1,078	1,152	1,264	770	520
Dairy products	419	329	484	407	521	465	481	301	211
Fruits and vegetables	633	548	680	560	716	702	793	457	348
Other food at home	$1,136	$906	$1,297	$1,120	$1,353	$1,322	$1,334	$904	$610
Food away from home	2,677	2,380	2,888	2,020	3,090	3,129	2,891	1,584	1,520
Alcoholic beverages	337	388	299	261	322	276	309	144	321
Housing	14,790	12,965	16,348	17,170	16,408	15,739	14,381	10,103	8,971
Shelter	8,412	7,271	9,415	10,076	9,470	8,861	7,981	5,873	5,450
Owned dwellings	6,406	5,336	7,380	7,565	7,567	6,890	5,784	2,646	2,464
Rented dwellings	1,370	1,195	1,460	2,244	1,418	1,012	1,707	3,088	2,688
Other lodging	637	740	575	267	485	959	490	139	298
Utilities, fuels, and public services	2,860	2,585	3,015	2,570	3,009	3,326	3,312	2,194	1,805
Natural gas	322	290	337	293	342	359	384	248	210
Housekeeping supplies	676	622	727	535	770	793	632	356	310
Household furnishings and equipment	1,994	1,953	2,066	1,848	2,093	2,201	1,739	984	971
Apparel and services	2,169	1,680	2,520	2,078	2,696	2,496	2,517	1,946	1,202
Transportation	9,289	8,067	10,214	9,368	9,585	12,029	9,785	4,694	4,521
Vehicle purchases (net outlay)	4,421	3,791	4,946	4,855	4,727	5,435	4,372	2,260	2,072
Gasoline and motor oil	1,376	1,146	1,522	1,281	1,454	1,819	1,643	720	705
Other vehicle expenses	$2,977	$2,549	$3,289	$2,904	$2,988	$4,153	$3,223	$1,484	$1,469
Public transportation	515	580	457	328	417	621	547	230	275
Health care	2,522	2,908	2,200	1,705	2,154	2,630	2,553	1,003	1,401
Entertainment	2,519	2,276	2,784	2,111	3,141	2,549	2,095	1,367	1,193
Personal care products and services	506	471	541	450	552	597	460	362	299
Reading	201	215	197	160	201	214	150	71	121
Education	829	528	1,115	317	1,030	1,822	582	426	424
Tobacco products and smoking supplies	324	269	342	239	344	406	500	239	279
Miscellaneous	1,042	931	1,091	1,094	1,001	1,236	1,321	843	707
Cash contributions	1,477	1,816	1,201	808	1,206	1,458	1,462	368	953
Personal insurance and pensions	4,812	4,291	5,301	4,951	5,382	5,380	4,413	1,827	1,962
Life and other personal insurance	602	548	644	398	703	693	616	170	170
Pensions and Social Security	4,210	3,743	4,658	4,553	4,679	4,687	3,798	1,657	1,792

[1] Components of income and taxes are derived from "complete income reporters" only.
[2] Value less than 0.05.
n.a. Not applicable.

SOURCE: "Table 5. Composition of consumer unit: Average annual expenditures and characteristics, Consumer Expenditure Survey, 1999," Bureau of Labor Statistics, Washington, DC, 2000

amount of tobacco production per acre, although the number of tobacco farms dropped. The number of farms growing tobacco in the United States decreased from about 512,000 in 1954 to approximately 89,706 in 1997. Similarly, the number of acres used to grow tobacco fell from 1.5 million acres in 1954 to 644,250 acres in 1999. The 1999 harvest produced more than 1.2 billion pounds of tobacco valued at about $2.3 billion. (See Table 7.3.)

In 1999 North Carolina led in tobacco production, followed by Kentucky, Tennessee, Virginia, South Carolina, and Georgia. (See Table 7.4.) For the leading half-dozen tobacco-producing states, tobacco plays a major role in the agricultural economy.

TABLE 7.3

Tobacco area, yield, production, and value, 1990–99

Year	Area harvested	Yield per acre	Production[1]	Marketing year average price per pound received by farmers	Value of production
	Acres	Pounds	1,000 pounds	Dollars	1,000 dollars
1990	733,310	2,218	1,626,380	1.738	2,827,167
1991	763,680	2,179	1,664,372	1.771	2,947,309
1992	784,440	2,195	1,721,671	1.777	3,059,246
1993	746,405	2,161	1,613,319	1.754	2,829,161
1994	671,065	2,359	1,582,896	1.758	2,779,056
1995	663,525	1,914	1,269,910	1.820	2,307,168
1996	733,060	2,072	1,518,704	1.882	2,853,739
1997	836,230	2,137	1,787,399	1.802	3,217,176
1998	717,605	2,062	1,479,867	1.828	2,700,795
1999	644,250	1,980	1,275,438	1.831	2,329,397

[1] Production figures are on farm-sales-weight basis.

SOURCE: "Table 2-44—Tobacco: Area, yield, production, and value, United States, 1990–99," in *Agricultural Statistics 2000,* National Agricultural Statistics Service, U.S. Department of Agriculture, Washington, D.C., 2000

TABLE 7.4

Tobacco area, yield, and production by state, 1997–99

State	Area harvested			Yield per harvested acre			Production		
	1997	1998	1999[1]	1997	1998	1999[1]	1997	1998	1999[1]
	Acres	Acres	Acres	Pounds	Pounds	Pounds	1,000 pounds	1,000 pounds	1,000 pounds
CT	2,545	2,815	2,950	1,622	1,519	1,709	4,128	4,276	5,042
FL	7,300	6,800	6,000	2,610	2,515	2,550	19,053	17,102	15,300
GA	43,000	41,000	33,000	2,075	2,200	1,940	89,225	90,200	64,020
IN	8,900	8,500	6,500	2,100	2,000	1,800	18,690	17,000	11,700
KY	250,500	226,260	221,700	1,988	1,961	1,826	497,928	443,628	404,863
MD	8,000	6,500	6,500	1,500	1,400	1,400	12,000	9,100	9,100
MA	1,175	1,265	1,310	1,628	1,413	1,731	1,913	1,788	2,267
MO	3,000	2,700	2,300	2,345	2,130	2,000	7,035	5,751	4,600
NC	321,400	251,100	208,200	2,275	2,197	2,160	731,199	551,730	449,620
OH	11,400	9,800	9,800	1,950	1,830	1,740	22,230	17,934	17,052
PA	8,100	7,800	6,200	2,100	2,015	1,802	17,020	15,720	11,170
SC	54,000	45,000	39,000	2,340	2,050	2,000	126,360	92,250	78,000
TN	59,480	59,415	59,270	1,922	1,870	1,866	114,292	111,100	110,569
VA	53,080	45,000	38,600	2,215	2,131	2,259	117,576	95,898	87,185
WV	1,800	1,600	1,600	1,700	1,350	1,350	3,060	2,160	2,160
WI	2,550	2,050	1,320	2,231	2,063	2,114	5,690	4,230	2,790
US	836,230	717,605	644,250	2,137	2,062	1,980	1,787,399	1,479,867	1,275,438

[1] Preliminary.

SOURCE: "Table 2-45—Tobacco: Area, yield, and production, by States, 1997–99," in *Agricultural Statistics 2000,* National Agricultural Statistics Service, U.S. Department of Agriculture, Washington, D.C., 2000

Every five years, the U.S. Bureau of the Census performs a census of farming and agriculture. In 1997 the national average value of tobacco sold per farm was $32,700. The value per farm varied widely from state to state, from $121,600 in Georgia to just $12,619 in Tennessee.

The 1997 census reported that almost all tobacco growers (97.6 percent) were white and the average age of the farmers was 53.5 years. The 55-and-older group is increasing, while the number of growers under age 55 is declining. This trend indicates that many younger people are choosing not to become tobacco farmers.

Value and Income

In 1998 tobacco was the sixth-largest cash crop, behind corn, soybeans, hay, wheat, and cotton. Preliminary USDA figures about the 2000 tobacco crop, worth more than $1.7 billion, indicate it represented 0.92 per-

TABLE 7.5

Cash receipts from farm marketings and tobacco, 1991–2000

Period	Cash receipts[3]				Tobacco as a percentage of	
	Livestock and products	All crops	Total farm	Tobacco	All Crops	Total cash receipts
	Million dollars				Percent	
1991	85,750	82,001	167,751	2,881	3.51	1.72
1992	85,596	85,662	171,346	2,962	3.46	1.73
1993	90,036	87,102	177,617	2,949	3.39	1.66
1994	88,107	91,562	180,775	2,645	2.89	1.46
1995	87,004	100,700	187,704	2,548	2.53	1.36
1996	93,005	106,575	199,579	2,796	2.62	1.40
1997	96,568	112,097	208,664	2,886	2.57	1.38
1998[1]	92,972	102,542	195,514	3,049	2.97	1.56
1999[1]	95,463	93,146	188,610	2,273	2.44	1.21
2000[2]	97,987	93,015	191,002	1,764	1.90	0.92

[1]Revised.   [2]Preliminary.   [3]Does not include government payments. Calendar year sales.

SOURCE: "Table 30: Cash receipts from farm marketings and tobacco, 1991–2000," in *Tobacco Situation and Outlook Report*, Economic Research Service, U.S. Department of Agriculture, Washington, DC, April 2001

TABLE 7.6

U.S. cigarette output, removals, and consumption, 1989–2000

Year	Total output	Taxable removals	Overseas forces and shipments 1/	Exports	Total U.S. consumption 2/
			Billion pieces		
1989	677.2	525.8	7.3	141.8	540.0
1990	709.7	523.2	14.5	164.3	525.0
1991	694.5	497.1	14.5	179.2	510.0
1992	718.5	509.4	7.4	205.6	500.0
1993	661.0	463.4	6.5	195.5	485.0
1994	725.5	488.6	11.4	220.2	486.0
1995	746.5	490.3	19.8	231.1	487.0
1996	754.5	486.0	17.1	243.9	487.0
1997	719.6	471.4	15.0	217.0	480.0
1998	679.7	457.9	11.2	201.3	465.0
1999	606.6	429.8	14.1	151.4	435.0
2000 3/	610.0	430.0	13.5	145.5	430.0

1/ To Puerto Rico and other U.S. possessions. Also includes ship stores and small tax-exempt categories.
2/ Allows for estimated inventory change for 1989 through 1999.
3/ Estimated.
Compiled from reports of the Bureau of Alcohol, Tobacco, and Firearms; Bureau of the Census; and The Tobacco Institute.

SOURCE: "Table 1—Cigarettes: U.S. output, removals, and consumption, 1950–2000," in *Tobacco Situation and Outlook Yearbook,* Economic Research Service, U.S. Department of Agriculture, Washington, D.C., December, 2000.

cent of all cash receipts from crops and nearly 2 percent of all farm commodities. (See Table 7.5.)

Tobacco is a very labor-intensive crop to produce; it takes 250 person-hours per acre to grow and harvest tobacco, compared with three person-hours for wheat. However, if the cost in person-hours is high, the monetary rewards to tobacco growers are equally high. In 1994 the value per acre for tobacco was $4,220 in gross income to the grower, far more than most other crops. Corn, for example, returned $304 per acre, and soybeans brought in $226.

Tobacco pays more per acre than any other commonly grown farm crop, but about 56 percent of tobacco growers cannot earn enough as tobacco farmers to make a living and must also work at other jobs, a common situation for all types of farming.

Manufacturing

The Economic Research Service of the U.S. Department of Agriculture (USDA) estimates that, in 2000, American factories produced 610 billion cigarettes, of which 145.5 billion were shipped to other countries. (See Table 7.6.) Four major companies produce cigarettes for the American market: Philip Morris, 49.1 percent market share in 1997; R.J. Reynolds, 24.2 percent; Brown and Williamson, 16.1 percent; and Lorillard, 8.7 percent. In 1997 more than 1.5 million retail outlets distributed tobacco products, including more than 560,000 vending machines.

TABLE 7.7

Per capita consumption of tobacco products (including overseas forces), 1991–2000

Year	Per capita 16 years and over	Per capita 18 years and over				Per male 18 years and over			
		Cigarettes[1]	Snuff[2]		All tobacco products[3]	Large cigars & cigarillos		Smoking tobacco[2]	Chewing tobacco[2]
	Number	Number	Pounds			Number		Pounds	
1991	2,737	2,834	4.9	0.3	5.6	26.4	0.43	0.20	0.80
1992	2,555	2,647	4.6	0.2	5.6	24.5	0.40	0.12	1.04
1993	2,453	2,543	4.7	0.2	5.7	23.4	0.38	0.11	1.01
1994	2,435	2,524	4.2	0.2	5.2	25.3	0.41	0.11	0.99
1995	2,417	2,505	4.2	0.2	5.1	27.5	0.45	0.09	0.97
1996	2,391	2,484	4.2	0.2	5.2	32.4	0.53	0.08	0.96
1997	2,331	2,422	4.1	0.2	5.1	37.3	0.61	0.08	0.93
1998	2,233	2,321	3.7	0.2	4.7	38.0	0.62	0.08	0.91
1999	2,067	2,148	3.6	0.2	4.6	39.5	0.65	0.09	0.90
2000[4]	2,025	2,103	3.5	0.2	4.5	39.1	0.64	0.09	0.90

[1]Unstemmed processing weight. [2]Finished product weight. [3]Includes small cigars. [4]Preliminary.

SOURCE: "Table 2: Per capita consumption of tobacco products in the United States (including overseas forces), 1991–2000," in *Tobacco Situation and Outlook Report*, Economic Research Service, U.S. Department of Agriculture, Washington, DC, April 2001

TABLE 7.8

Expenditures for tobacco products and disposable personal income, 1991–2000[1]

Year	Total	Cigarettes	Cigars[2]	Other[3]	Disposable personal income	Percent of disposable personal income spent on tobacco products			
						All	Cigarettes	Cigars[2]	Other[3]
	Million dollars				Billion dollars	Percent			
1991	45,305	42,850	705	1,840	4,231	1.08	1.02	0.02	0.04
1992	48,470	45,790	715	1,965	4,500	1.08	1.02	0.02	0.04
1993	48,955	46,150	730	2,075	4,789	1.04	0.98	0.02	0.04
1994	47,297	44,544	766	1,987	5,022	0.96	0.90	0.02	0.04
1995	48,692	45,793	846	2,053	5,356	0.92	0.86	0.02	0.04
1996	50,223	47,233	872	2,118	5,535	0.90	0.85	0.02	0.04
1997	52,569	49,437	915	2,217	5,795	0.91	0.85	0.02	0.04
1998	56,024	51,987	1,607	2,430	6,320	0.98	0.92	0.02	0.04
1999[4]	70,641	66,286	1,788	2,567	6,638	1.06	1.00	0.03	0.04
2000[5]	77,496	72,945	1,853	2,698	6,989	1.11	1.04	0.03	0.04

[1]Expenditures exclude sales tax. [2]Includes small cigars (cigarette-size). [3]Smoking tobacco, chewing tobacco, and snuff. [4]Subject to revision. [5]Estimated.

SOURCE: "Table 32: Expenditures for tobacco products and disposable personal income, 1991–2000," in *Tobacco Situation and Outlook Report*, Economic Research Service, U.S. Department of Agriculture, Washington, DC, April 2001

Tobacco Consumption

In 2000 U.S. tobacco consumption, including that of overseas armed forces, was 430 billion cigarettes, 3.8 billion cigars, and 2.3 million cigarillos (small, thin cigars). These figures do not include pipe and roll-your-own tobacco, chewing tobacco, and snuff.

Cigarette smoking in the United States has been dropping almost every year since 1963, when per capita consumption reached a record high of 4,345 cigarettes. In 2000 preliminary USDA figures estimated the per capita cigarette consumption of the population 18 years old and older at 2,103 cigarettes. (See Table 7.7.)

Per capita consumption of snuff by those 18 and over remained the same from 1992 to 2000. However, the number of cigarettes consumed per capita during that time span decreased. The number of cigars and cigarillos con-sumed per male 18 years and older increased by 54 percent from 1992 to 2000, attesting to the current popularity of cigar clubs. (See Table 7.7.) The overall decline in tobacco consumption has been caused by a variety of factors, including higher prices, antismoking campaigns, and further bans on cigarette smoking in many public places.

Virtually all cigarettes sold in 2000 (98.7 percent) were filter tip. Economy brands' share of the market rose to a high of 37 percent in 1993, but fell to 26.7 percent in 2000.

Consumer Spending on Tobacco

Americans spent an estimated $77.5 billion on tobacco products in 2000, the highest amount in the past decade. Ninety-four percent ($72.9 billion) was spent on cigarettes. This was slightly more than 1 percent of all disposable personal income. (See Table 7.8.) Although consumption of cigarettes has declined over the past few

TABLE 7.9

U.S. cigarette exports to leading destinations, 1995–2000

						January - September	
Country	1995	1996	1997	1998	1999	1999	2000 1/
				Billion pieces			
European Union	85.2	71.1	50.0	48.9	20.3	17.6	9.3
Japan	61.7	67.7	67.7	70.9	72.5	54.7	57.9
Lebanon	10.1	11.7	10.3	10.9	5.7	4.5	3.6
Saudi Arabia	9.7	9.3	9.3	8.7	10.2	7.3	8.2
Singapore	7.5	8.0	10.3	7.3	3.3	2.5	2.3
Cyprus	7.5	9.5	0.8	7.0	6.0	5.0	5.0
Hong Kong	6.7	6.1	5.9	5.8	3.2	2.3	2.3
South Korea	6.6	6.3	9.9	5.3	2.8	1.8	3.3
Israel	3.7	3.2	3.2	4.0	4.4	3.3	3.4
Russia	3.7	17.0	4.3	3.7	0.8	0.3	1.3
United Arab Emirates	3.2	1.8	3.4	3.3	1.6	1.6	1.0
Paraguay	2.2	1.9	2.2	2.3	1.0	0.7	0.5
Kuwait	1.9	1.8	2.8	2.2	1.9	1.5	1.3
Morocco	1.9	1.6	1.8	1.8	1.4	1.2	0.2
Panama	1.8	2.4	1.9	1.8	0.5	0.3	0.2
Taiwan	1.6	2.2	1.8	1.8	1.9	1.5	1.3
Mexico	1.3	2.1	7.2	1.2	0.3	0.1	0.2
Turkey	*	2.7	2.4	0.9	0.1	0.1	0.8
Netherlands Antilles	0.4	0.4	0.9	0.3	0.2	0.2	0.1
Other countries	14.4	17.1	20.2	62.1	33.6	26.8	18.1
Total	231.1	243.9	217.0	201.3	151.4	115.7	111.0

*Less than 50 million. 1/ Subject to revision.
Compiled from publications and records of the Bureau of the Census.

SOURCE: "Table 6—U.S. cigarette exports to leading destinations, 1995–2000," in *Tobacco Situation and Outlook Yearbook,* Economic Research Service, U.S. Department of Agriculture, Washington, D.C., December, 2000.

years, expenditures (after adjustment for inflation) have increased, owing to increasing prices.

Exports

As in previous years, the United States was a leading exporter of tobacco in 1998. U.S. exports of unmanufactured tobacco and tobacco products were valued at $6.3 billion.

In 1999 just over 150 billion cigarettes were exported to well over 100 countries. Most of the cigarettes exported went to Japan (72.5 billion) and the countries of the European Union (20.3 billion). Other major importers of American cigarettes were Saudi Arabia (10.2 billion), Cyprus (6 billion), Lebanon (5.7 billion), Israel (4.4 billion), and Taiwan (1.9 billion). (See Table 7.9.) Many of the importers in these countries then export these cigarettes to other countries.

Imports

The United States was also the leading importer of tobacco in 1999. About 491 million pounds of tobacco were imported in that year, falling sharply from 668 million pounds in 1996. Lower cigarette production decreased demand for foreign tobacco. Most came from Turkey and Brazil.

WORLD TOBACCO PRODUCTION

The U.S. Department of Agriculture estimated the world production of tobacco in 1999 at 6.8 million metric

tons (a metric ton equals 1,000 kilograms or 2,204.62 pounds). China produced about 35 percent (about 2.4 million metric tons). Other leading tobacco producers included the United States (635,029 metric tons), India (648,600 metric tons), Brazil (569,000 metric tons), Turkey (238,600 metric tons), Indonesia (210,000 metric tons), and Zimbabwe (192,025 metric tons). (See Table 7.10.)

ALCOHOL AND TOBACCO ADVERTISING

Alcohol Advertising

According to the Center for Science in the Public Interest, the alcohol industry spends nearly $3 billion a year on marketing and promotion. In 1999 alcohol advertising expenditures accounted for approximately $1.24 billion of that total. Broadcast alcohol advertising cost about $795.5 million, 64 percent of the total spending on alcohol advertising. Most of that amount (88 percent) was spent on beer advertising. Expenditures for broadcast advertising on distilled spirits totaled just over $17 million in 1999, a twelvefold increase from 1995. This huge increase reflects the lifting of the voluntary ban on radio and television advertising of distilled spirits.

LIFTING THE BAN. The distilled spirits industry, in a self-imposed ban, had not advertised on television and radio for 50 years. In November 1996, however, the Distilled Spirits Council of the United States (DISCUS) announced that it had revised its advertising code and lift-

TABLE 7.10

Tobacco area, yield, and production worldwide by country, 1997–99

Continent and country	Area harvested			Yield per hectare			Production[2]		
	1997	1998	1999[1]	1997	1998	1999[1]	1997	1998	1999[1]
	Hectares	Hectares	Hectares	Metric tons	Metric tons	Metric tons	Metric tons	Metric tons	Metric tons
North America:									
Canada	28,500	27,800	25,091	2.50	2.49	2.59	71,110	69,300	64,864
Mexico	25,385	31,808	27,709	1.74	1.93	1.89	44,293	61,457	52,333
United States	328,406	302,324	262,184	2.47	2.30	2.42	810,154	696,116	635,029
Total	382,291	361,932	314,984	2.42	2.28	2.39	925,557	826,873	752,226
South America:									
Argentina	71,000	79,000	69,000	1.74	1.47	1.63	123,200	116,500	112,500
Bolivia	1,250	1,250	1,250	1.00	1.00	1.00	1,250	1,250	1,250
Brazil	329,500	346,000	338,000	1.75	1.29	1.68	576,600	447,000	569,000
Chile	3,499	3,987	4,089	3.08	2.85	2.95	10,772	11,374	12,049
Colombia	17,905	17,905	17,905	1.57	1.57	1.57	28,178	28,178	28,178
Ecuador	1,000	1,000	1,000	3.85	3.85	3.85	3,850	3,850	3,850
Guyana	100	100	100	1.00	1.00	1.00	100	100	100
Paraguay	5,200	5,200	5,200	1.75	1.75	1.75	9,100	9,100	9,100
Peru	2,500	2,500	2,500	1.24	1.24	1.24	3,100	3,100	3,100
Uruguay	800	800	800	1.75	1.75	1.75	1,400	1,400	1,400
Venezuela	7,328	8,500	8,500	2.50	2.47	2.54	18,329	21,000	21,550
Total	440,082	466,242	448,344	1.76	1.38	1.70	775,879	642,852	762,077
Central America:									
Costa Rica	1,072	1,072	1,072	2.03	2.03	2.03	2,180	2,180	2,180
El Salvador	561	580	580	1.85	1.79	1.79	1,038	1,038	1,038
Guatemala	8,275	8,873	7,637	2.24	2.30	2.28	18,515	20,440	17,412
Honduras	5,157	5,157	5,157	1.78	1.78	1.78	9,177	9,177	9,177
Nicaragua	2,240	2,240	2,240	2.03	2.03	2.03	4,550	4,550	4,550
Panama	1,094	1,094	1,094	2.00	2.00	2.00	2,188	2,188	2,188
Total	18,399	19,016	17,780	2.05	2.08	2.06	37,648	39,573	36,545
Caribbean:									
Cuba	59,000	59,000	59,000	0.63	0.63	0.63	37,000	37,000	37,000
Dominican Rep	21,171	27,050	25,698	1.43	1.51	1.35	30,279	40,950	34,808
Haiti	565	565	565	1.29	1.29	1.29	730	730	730
Jamaica	1,175	1,175	1,175	1.99	1.99	1.99	2,339	2,339	2,339
St. Vincent	70	70	70	1.21	1.21	1.21	85	85	85
Trinidad & Tob.	100	100	100	1.70	1.70	1.70	170	170	170
Total	82,081	87,960	86,608	0.86	0.92	0.87	70,603	81,274	75,132
European Union:									
Austria	103	101	102	1.93	2.03	1.72	199	205	175
Belgium-Lux	322	320	320	3.98	3.69	3.69	1,280	1,180	1,180
France	9,079	9,067	9,040	2.61	2.92	2.91	23,662	26,513	26,348
Germany	3,501	3,831	3,990	2.43	2.40	2.45	8,504	9,200	9,770
Greece	67,250	67,230	67,200	1.97	1.97	1.94	132,450	132,200	130,500
Italy	48,120	48,600	48,600	2.92	2.82	2.73	140,634	136,944	132,500
Portugal	2,909	2,909	2,909	2.14	2.14	2.14	6,226	6,226	6,226
Spain	13,225	13,300	13,250	3.20	3.18	3.19	42,290	42,300	42,300
Total	144,509	145,358	145,411	2.46	2.44	2.40	355,245	354,768	348,999
Western Europe:									
Switzerland	655	635	635	1.59	2.09	2.09	1,039	1,325	1,325
Eastern Europe:									
Albania	8,000	8,500	8,700	0.89	0.92	0.92	7,110	7,830	8,010
Bulgaria	48,511	34,400	33,800	1.62	1.31	1.45	78,510	45,107	48,912
Croatia	7,274	7,837	7,700	1.81	1.73	1.74	13,154	13,523	13,398
Hungary	6,600	7,000	7,300	1.64	2.20	2.10	10,800	15,375	15,300
Macedonia (Skopje)	22,000	22,000	22,000	1.36	1.36	1.36	30,000	30,000	30,000
Poland	19,040	19,675	19,800	2.26	2.19	2.17	42,990	43,000	43,000
Romania	11,560	12,000	12,000	1.21	1.23	1.23	13,980	14,750	14,750
Slovakia	2,000	2,000	2,000	1.75	1.75	1.75	3,500	3,500	3,500
Yugoslavia	5,150	6,200	6,200	1.41	1.23	1.23	7,271	7,604	7,604
Total	130,135	119,612	119,500	1.59	1.51	1.54	207,315	180,689	184,474

ed the voluntary ban. DISCUS president and CEO Fred A. Meister said, "The absence of spirits from television and radio has contributed to the mistaken perception that spirits are somehow 'harder' or worse than beer or wine and thus deserving of harsher social, political, and legal treatment." Meister continued, "Distilled spirits advertisement will continue to be responsible, dignified, and tasteful messages for adults and will avoid targeting those under the legal purchase age, regardless of the medium."

REACTIONS TO LIFTING THE BAN. Reactions to the announcement were mixed. President Bill Clinton called

TABLE 7.10

Tobacco area, yield, and production worldwide by country, 1997–99 [CONTINUED]

Continent and country	Area harvested			Yield per hectare			Production[2]		
	1997	1998	1999[1]	1997	1998	1999[1]	1997	1998	1999[1]
	Hec-tares	Hec-tares	Hec-tares	Metric tons	Metric tons	Metric tons	Metric tons	Metric tons	Metric tons
FSU-12:[3]									
Armenia	4,304	4,304	4,304	0.92	0.92	0.92	3,966	3,966	3,966
Azerbaijan	4,500	4,500	4,500	2.33	2.33	2.33	10,500	10,500	10,500
Belarus	1,076	1,076	1,076	2.42	2.42	2.42	2,606	2,606	2,606
Georgia	5,380	5,400	5,400	1.64	1.63	1.63	8,800	8,800	8,800
Kazakhstan	1,900	5,500	4,000	2.01	1.82	2.00	3,820	10,000	8,000
Kyrgyzstan	12,000	12,000	12,000	2.50	2.50	2.50	30,000	30,000	30,000
Moldova	20,000	25,000	28,000	1.41	1.22	1.16	28,260	30,434	32,608
Russia	700	990	900	0.91	0.53	0.67	640	520	600
Tajikistan	3,228	3,228	3,228	3.28	2.48	2.48	10,593	8,000	8,000
Turkmenistan	1,076	1,100	1,100	2.39	2.36	2.36	2,570	2,600	2,600
Ukranine	5,600	5,000	5,000	0.80	0.84	0.84	4,500	4,200	4,200
Uzbekistan	8,500	9,000	6,700	2.74	3.11	3.43	23,300	28,000	23,000
Total	68,264	77,098	76,208	1.90	1.81	1.77	129,555	139,626	134,880
North Africa:									
Algeria	2,700	2,700	2,700	1.96	1.96	1.96	5,300	5,300	5,300
Libya	900	900	900	1.61	1.61	1.61	1,450	1,450	1,450
Morocco	3,500	3,500	3,500	1.13	1.13	1.13	3,962	3,962	3,962
Tunisia	6,700	6,700	6,700	1.18	1.18	1.18	7,900	7,900	7,900
Total	13,800	13,800	13,800	1.35	1.35	1.35	18,612	18,612	18,612
Other Africa:									
Angola	3,950	3,950	3,950	0.99	0.99	0.99	3,900	3,900	3,900
Benin	200	200	200	2.00	2.00	2.00	400	400	400
Burundi	705	705	705	1.00	1.00	1.00	705	705	705
Cameroon	3,400	3,400	3,400	1.62	1.62	1.62	5,500	5,500	5,500
Central African Rep	750	750	750	0.87	0.87	0.87	650	650	650
Chad	200	200	200	1.00	1.00	1.00	200	200	200
Congo (Brazzaville)	4,000	4,000	4,000	0.45	0.45	0.45	1,800	1,800	1,800
Cote d'Ivoire	10,000	10,000	10,000	0.26	0.26	0.26	2,600	2,600	2,600
Ethiopia	3,000	3,000	3,000	1.17	1.17	1.17	3,500	3,500	3,500
Ghana	3,950	3,950	3,950	0.38	0.38	0.38	1,500	1,500	1,500
Kenya	8,805	8,805	8,805	2.51	2.51	2.51	22,120	22,120	22,120
Liberia	10	10	10	1.00	1.00	1.00	10	10	10
Madagascar	5,900	5,900	5,900	0.93	0.93	0.93	5,500	5,500	5,500
Malawi	122,300	104,200	100,200	1.30	1.30	1.27	158,615	135,138	127,150
Mali	1,000	1,000	1,000	0.55	0.55	0.55	550	550	550
Mauritius	655	655	655	1.63	1.63	1.63	1,065	1,065	1,065
Mozambique	2,700	2,700	2,700	1.07	1.07	1.07	2,900	2,900	2,900
Niger	1,000	1,000	1,000	0.93	0.93	0.93	930	930	930
Nigeria	10,000	10,000	10,000	2.30	2.10	2.10	23,000	21,000	21,000
Reunion	200	200	200	1.00	1.00	1.00	200	200	200
Sierra Leone	540	540	540	1.11	1.11	1.11	600	600	600
South Africa; Rep	14,905	14,905	14,064	1.81	2.20	2.32	27,000	32,768	32,683
Swaziland	200	200	200	1.00	1.00	1.00	200	200	200
Tanzania; United Rep	33,900	33,900	33,900	0.74	0.74	0.42	25,080	25,080	14,070
Togo	4,000	4,000	4,000	0.00	0.00	0.00	2,000	2,000	2,000
Uganda	7,525	7,525	7,525	0.96	0.96	0.96	7,198	7,198	7,198
Zaire	3,700	3,700	3,700	1.11	1.11	1.11	4,110	4,110	4,110
Zambia	4,882	4,882	4,882	1.29	1.29	1.29	6,300	6,300	6,300
Zimbabwe	97,750	99,566	91,910	1.97	2.25	2.09	192,144	223,977	192,025
Total	350,127	333,843	321,346	1.43	1.53	1.45	500,277	512,401	461,366

the ending of the ban a "simply irresponsible move" that would make the job of rearing children harder. In April 1997 he called for an inquiry by the Federal Communications Commission (FCC) into the impact of lifting the ban.

The president did not call for a similar study into the effects of beer and wine advertising, although he left the door open for broadening the study. Karolyn Nunnallee, president of Mothers Against Drunk Driving (MADD), questioned the omission of beer and wine advertising. "When beer is the No. 1 alcoholic beverage of choice among our youth," she said, "it just doesn't make sense that these beer ads would not be targeted also."

More than 240 health, consumer, safety, and religious organizations submitted a petition to the FCC requesting an inquiry into the effects of broadcast alcohol advertisements. The FCC commissioners had a mixed response to the request for a study. FCC chairman Reed Hundt supported the request for an official study, while commissioners Rachelle Chong and James Quello suggested the FCC has no jurisdiction over advertising and that the Federal Trade Commission (FTC) would be the proper agency to consider the study. The vote was tied, automatically denying the request for a study.

In 1998, prompted by a congressional order, the FTC investigated whether the alcohol industry was doing

TABLE 7.10

Tobacco area, yield, and production worldwide by country, 1997–99 [CONTINUED]

Continent and country	Area harvested			Yield per hectare			Production[2]		
	1997	1998	1999[1]	1997	1998	1999[1]	1997	1998	1999[1]
	Hec-tares	Hec-tares	Hec-tares	Metric tons	Metric tons	Metric tons	Metric tons	Metric tons	Metric tons
Other Asia:									
Bangladesh	50,263	50,263	50,263	0.88	0.88	0.88	44,000	44,000	44,000
Burma	36,000	36,000	36,000	1.22	1.22	1.22	44,000	44,000	44,000
Cambodia	9,000	9,000	9,000	0.56	0.56	0.56	5,000	5,000	5,000
China	2,353,000	1,647,000	1,310,000	1.81	1.80	1.82	4,251,000	2,966,000	2,380,000
India	420,200	432,780	429,940	1.48	1.46	1.51	623,700	633,200	648,600
Indonesia	216,500	214,000	212,000	0.08	0.94	0.99	17,500	202,000	210,000
Japan	25,662	26,214	25,725	2.67	2.56	2.49	68,504	67,100	63,960
Korea, North	20,000	20,000	20,000	1.33	1.33	1.33	26,640	26,640	26,640
Korea, South	27,181	25,730	25,730	2.00	2.02	2.02	54,388	52,040	52,040
Laos	4,000	4,000	4,000	0.75	0.75	0.75	3,000	3,000	3,000
Malaysia	11,297	14,720	13,000	1.06	0.80	0.81	11,965	11,805	10,540
Pakistan	45,862	50,317	53,100	1.88	1.84	1.85	86,279	92,728	98,500
Philippines	29,397	45,171	41,011	2.07	1.61	1.43	60,900	72,670	58,500
Sri Lanka	12,165	12,165	12,165	0.74	0.74	0.74	9,000	9,000	9,000
Taiwan	4,061	4,321	4,300	2.53	2.34	2.35	10,283	10,120	10,120
Thailand	47,000	51,000	51,800	1.47	1.33	1.18	69,250	67,600	61,000
Vietnam	36,000	36,000	36,000	0.89	0.89	0.89	32,000	32,000	32,000
Total	3,347,588	2,678,681	2,334,034	1.62	2.00	2.00	5,417,409	4,338,903	3,756,900
Middle East:									
Cyprus	161	161	161	1.50	1.50	1.50	241	241	241
Iran	18,000	18,000	18,000	1.39	1.39	1.39	25,000	25,000	25,000
Iraq	2,000	2,000	2,000	1.09	1.09	1.09	2,180	2,180	2,180
Jordan	2,100	2,100	2,100	1.27	1.29	1.29	2,668	2,700	2,700
Lebanon	3,750	3,750	3,750	1.33	1.33	1.33	5,000	5,000	5,000
Oman	1,800	1,800	1,800	1.11	1.11	1.11	2,000	2,000	2,000
Syria	15,000	15,000	15,000	1.15	1.15	1.15	17,208	17,200	17,200
Turkey	323,000	288,300	263,600	0.96	0.90	0.91	310,850	260,750	238,600
United Arab. Emirates	350	350	350	5.71	5.71	5.71	2,000	2,000	2,000
Yemen	3,300	3,300	3,300	1.73	1.73	1.73	5,720	5,720	5,720
Total	369,461	334,761	310,061	1.01	0.96	0.97	372,867	322,791	300,641
Oceania:									
Australia	3,200	2,900	2,800	2.88	3.10	2.68	9,200	9,000	7,500
New Zealand	600	600	600	2.58	2.58	2.58	1,550	1,550	1,550
Solomon Islands	100	100	100	0.95	0.95	0.95	95	95	95
Total	3,900	3,600	3,500	2.78	2.96	2.61	10,845	10,645	9,145
World Total	5,351,292	4,642,538	4,192,201	1.65	1.61	1.63	8,822,851	7,470,332	6,842,322

[1] Preliminary.
[2] Production data in metric tons, on farm-sales-weight basis, which is about 10 percent above dry-weight data normally reported in trade statistics.
[3] FSU–12 includes the 12 newly independent States of the former USSR.

Prepared or estimated on the basis of official statistics of foreign governments, other foreign source materials, reports of U.S. Agricultural Counselors, Attaches, Foreign Service Officers and results of office research, and related information.

SOURCE: "Table 2-46—Tobacco: Area, yield, and production in specified countries, 1997–99," in *Agricultural Statistics 2000*, National Agricultural Statistics Service, U.S. Department of Agriculture, Washington, D.C., 2000

enough to discourage ads that appeal to underage drinkers. In a report to Congress in September 1999, the FTC called for beer, wine, and liquor companies to do more to ensure that minors do not see their advertisements. The FTC recommended that ads for alcoholic beverages should not be shown in theaters with movies rated PG or PG-13, on TV programs aimed at similar audiences, or on college campuses. Additionally, the FTC called for the formation of independent review boards to handle complaints from consumers and competitors about alcohol advertisements.

PUBLIC SUPPORT FOR RESTRICTIONS ON ALCOHOL ADVERTISING. In its report, the FTC did not recommend government regulation of alcohol advertising. Proponents of alcohol advertising restrictions, however, believe that industry self-regulation has failed and should be replaced by stricter government regulations.

The *Youth Access to Alcohol Survey* (Alcohol Epidemiology Program, University of Minnesota, 1998), released by the Robert Wood Johnson Foundation, found that Americans are very concerned about underage drinking and support restrictions on alcohol advertising. Two-thirds (66.5 percent) favored a ban on hard-liquor advertising on television (see Figure 7.1), and 60.7 percent said they supported a ban on beer and wine ads on television (see Figure 7.2). Approximately 59 percent favored a law that would prohibit the use of sports teams and athletes in alcohol ads (see Figure 7.3).

According to the Center for Science in the Public Interest (CSPI), a nonprofit health-advocacy organization based in Washington, D.C., children frequently exposed to television beer ads can usually match brand names and beer slogans. More recognize the Budweiser frogs than recognize Tony the Tiger, Smokey the Bear, or Mighty Morphin' Power Rangers. Mothers Against Drunk Driving (MADD) has singled out Budweiser and its brewer, Anheuser-Busch, for criticism. MADD president Karolyn Nunnallee charges that "campaigns such as the Budweiser lizards and frogs are unconscionable at a time when underage drinking is at epidemic levels. Alcohol marketers are bombarding our children with characters that look like they belong on Saturday morning cartoons, and it's absurd to think these don't affect our young people."

The CSPI observes that children frequently exposed to television beer advertising believe that "beer consumption is related to good times and fun rather than caution and risk." The alcohol industry denies that any link exists between their advertising and underage drinking. As proof, it points to long-term trends showing that alcohol consumption among youth has stabilized.

In 1998 the National Institute on Alcohol Abuse and Alcoholism (NIAAA) and the federal Center for Substance Abuse Prevention (CSAP) began a five-year longitudinal joint research study to determine whether alcohol advertising affects the drinking behavior of youth.

Tobacco Advertising

In 1996 the tobacco industry spent $243 million dollars on magazine advertising for smoking products. (Smoking products may not be advertised on television.) In the past, cigarette advertisements in magazines were usually full-page and full-color productions showing glamorous, sophisticated people or rugged, outdoor types, depending on the market toward which the advertisement was targeted. Food and Drug Administration rules that were to become effective in August 1996 would have permitted only black and white, text-only advertising. However, lawsuits were filed against the FDA, contending, among other challenges, that these advertising restrictions violated the First Amendment protection of free commercial speech.

In March 2000 the Supreme Court ruled that federal health authorities, such as the FDA, have no power to regulate the manufacture and sale of cigarettes. A few days after the ruling was announced, President Bill Clinton urged Congress to give the FDA the authority to regulate tobacco advertising and to make illegal tobacco purchases by children under 18. In 2001 a bipartisan coalition renewed the effort to grant the Food and Drug Administration authority to regulate tobacco, reintroducing a bill that would let the FDA restrict tobacco advertising that targets children. The proposed legislation is called "A Bill to Provide for the Protection of Children from Tobacco,"

and has been referred to the Senate Committee on Health, Education, Labor, and Pensions.

ADVERTISING TARGETED TOWARD WOMEN. Women did not generally smoke before World War I (1914–18). Just after the war, Lorillard began using images of women to advertise its Murad and Helman brands. During the 1920s, Marlboro advertised in fashion magazines, and Lucky Strike encouraged women to "Reach for a Lucky Instead of a Sweet," implying they could stay slim by smoking.

In the late 1960s, Philip Morris introduced its successful Virginia Slims brand. Its "You've Come a Long Way, Baby" campaign implied that smoking cigarettes was as much a part of women's rights as equal pay. John P. Pierce et al., in "Smoking Initiation by Adolescent Girls, 1944 Through 1988" (*Journal of the American Medical Association*, February 23, 1994), found that "in girls younger than 18 years, smoking initiation increased abruptly around 1967, when tobacco advertising aimed at selling specific brands to women was introduced. This increase was particularly marked in those females who never attended college. Initiation rates for females younger than 18 years peaked around 1973, at about the same time sales of these brands peaked."

ADVERTISING TARGETED TOWARD TEENAGERS. Critics of tobacco advertising have frequently cited the cartoon character Joe Camel, which R.J. Reynolds used to advertise its Camel brand of cigarettes, as an example of advertising targeted to teenagers. A study reported in 1991 in the *Journal of the American Medical Association* claimed that young people had been strongly influenced by the Joe Camel campaign. Among children, Joe Camel was as well recognized as Mickey Mouse. Between 1987, when the campaign began, and 1993, the percentage of Camel smokers under the age of 18 increased from 3 percent to 16 percent.

The R.J. Reynolds Tobacco Company responded with a November 1993 survey showing that Joe Camel was no more recognizable to teenagers than any other advertising character. Furthermore, recognition of Joe Camel did not necessarily influence teenagers to start smoking. The tobacco industry maintains that it has no desire to see teenagers smoking. According to the industry, peer pressure, not advertising, is the major factor that leads young people to smoke. Cigarette advertising is designed to attract the 60 percent of smokers who change their brand of cigarettes at some time during their lifetime.

Nevertheless, the Federal Trade Commission (FTC) officially issued an administrative complaint in May 1997. The FTC charged that the Joe Camel advertising campaign violated federal fair trade laws by promoting a lethal and addictive product to children and adolescents who could not legally purchase or use it. In 1997 the R.J. Reynolds Tobacco Company agreed to eliminate Joe

FIGURE 7.1

Opinions of respondents on whether they would favor a law that would ban all advertisement of hard liquor on television, 1998

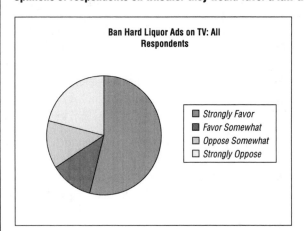

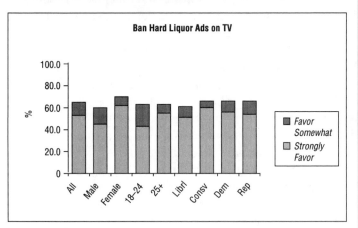

% in Favor of Banning Hard Liquor on TV [1]								
	All	**Male**	**Female**	**18–24**	**25+**	**Liberal**	**Conservative**	**Democrat**	**Republican**
Strongly Favor	55.0	46.4	62.6	46.9	56.2	51.7	59.3	57.5	56.3
Favor Somewhat	11.5	13.6	9.6	20.2	10.2	11.9	8.7	10.6	12.0
Oppose Somewhat	12.2	15.1	9.6	14.9	11.8	12.8	12.5	9.6	15.1
Strongly Oppose	21.4	25.0	18.2	18.1	21.8	23.6	19.5	22.2	16.6

[1] Weighted percents are calulated on valid responses only (missing or refused are not included) and may not add up to 100% due to rounding.

SOURCE: E.M. Harwood, A.C. Wagenaar, and K.M. Zander, *Youth Access to Alcohol Survey: Summary Report*, University of Minnesota, Minneapolis, MN, 1998

Camel and other cartoon characters from their advertisements and packaging.

U.S. tobacco companies are in a bind. Between smokers who quit and those who die, the companies lose a few million smokers every year. Since 95 percent of smokers start by the age of 19, it would seem a smart business move to target marketing campaigns to younger people to protect future revenues. However, any such campaign would likely lead to an immediate response from the government, parents, antismoking advocacy groups, and community leaders who oppose the deliberate targeting of young people in tobacco and alcohol advertising.

Billboard Advertising

At one time, one in every three billboards carried a tobacco advertisement. But on April 23, 1999, all tobacco billboards in the United States had to come down as part of the settlement between the tobacco industry and state attorneys general in 46 states. The tobacco companies accepted the FDA's authority to regulate tobacco products and restrict their advertising. However, the outdoor advertising ban applied only to signs larger than 14 square feet. Retailers put up smaller cigarette signs in front of their stores, "right at the kids' eye level," stated William Godshall, executive director of Smokefree Pennsylvania.

In 1998 a federal judge struck down New York City's Youth Protection Against Tobacco Advertising and Promotion Act. This law prohibited tobacco advertisements on doors and awnings and in windows. The judge ruled that only the federal government (via the enacting of laws by Congress) could address public health concerns by imposing restrictions on tobacco advertising. Additionally, other court decisions in 1997, 1998, and 2000 ruled that the FDA did not have the authority to regulate tobacco products, which invalidated all of the FDA's 1996 regulations.

A number of cities and states then implemented restrictions on alcohol and tobacco billboard advertising. For example, the Los Angeles city council and the state of Massachusetts voted to ban tobacco and alcohol billboards and other outdoor advertising from locations within 1,000 feet of parks, schools, and residential areas. The tobacco industry challenged the Massachusetts ban based on the right to free speech. In January 2001 the U.S. Supreme Court agreed to hear this case, and in June 2001 ruled that efforts to ban tobacco advertising near playgrounds and schools violate federal law and free speech rights. The Court added that cities and states may not add restrictions to the federal law that bans cigarette advertising on television and requires warning labels on packages. Only Congress has the power to amend federal law.

FIGURE 7.2

Opinions of respondents on whether they would favor a law that would ban all advertisement of beer and wine on television, 1998

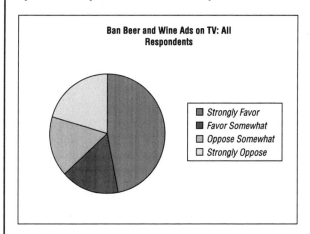

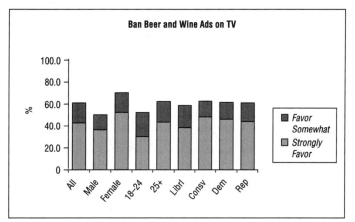

% in Favor of Banning Beer and Wine on TV [1]

	All	Male	Female	18–24	25+	Liberal	Conservative	Democrat	Republican
Strongly Favor	45.0	35.8	53.1	31.6	46.8	39.4	50.2	48.6	45.8
Favor Somewhat	15.7	15.8	15.6	20.5	15.2	17.0	13.9	13.8	15.5
Oppose Somewhat	18.4	22.7	14.8	26.6	17.3	21.7	14.6	18.0	19.3
Strongly Oppose	20.8	25.7	16.5	21.4	20.7	21.9	21.3	19.6	19.5

[1] Weighted percents are calulated on valid responses only (missing or refused are not included) and may not add up to 100% due to rounding.

SOURCE: E.M. Harwood, A.C. Wagenaar, and K.M. Zander, *Youth Access to Alcohol Survey: Summary Report*, University of Minnesota, Minneapolis, MN, 1998

FIGURE 7.3

Opinions of respondents on whether they would favor a law banning the use of sports teams and athletes as symbols in advertising and promotion of alcoholic beverages, 1998

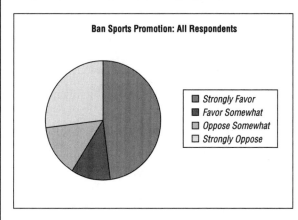

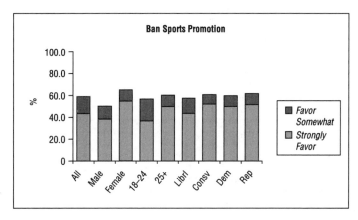

% in Favor of Banning Sports Promotion [1]

	All	Male	Female	18–24	25+	Liberal	Conservative	Democrat	Republican
Strongly Favor	48.2	39.7	55.6	36.1	49.9	44.3	52.0	49.4	51.1
Favor Somewhat	10.6	10.6	10.6	16.9	9.7	11.9	8.3	9.7	11.7
Oppose Somewhat	14.6	19.0	10.8	19.5	14.0	15.4	13.6	13.0	15.4
Strongly Oppose	26.6	30.7	23.0	27.5	26.5	28.5	26.1	27.8	21.9

[1] Weighted percents are calulated on valid responses only (missing or refused are not included) and may not add up to 100% due to rounding.

SOURCE: E.M. Harwood, A.C. Wagenaar, and K.M. Zander, *Youth Access to Alcohol Survey: Summary Report*, University of Minnesota, Minneapolis, MN, 1998

CHAPTER 8
THE GOVERNMENT AND THE COURTS

TAXATION

Taxation is an age-old method by which the government raises money. Alcoholic beverages have been taxed since colonial times, and tobacco products have been taxed since 1863. The alcohol and tobacco industries contribute a great deal of tax money to federal, state, and local governments.

Alcohol Taxes

According to the Distilled Spirits Council of the United States, Inc. (DISCUS), liquor is the most highly taxed consumer product in the nation. Direct and indirect local, state, and federal taxes account for 58 percent of the typical bottle price. While the beer and wine industries are taxed at lower levels, they also contribute a significant amount of tax revenue.

In fiscal year (FY) 1999, the federal government collected approximately $7.7 billion in excise taxes on alcoholic beverages. At $13.50 per proof gallon, federal distilled spirits taxes amounted to about $3.7 billion, which included taxes on both domestic and imported distilled spirits. Wine taxes depend on several variables, such as alcohol content and the size of the winery; in 1999 federal excise taxes on wine totaled $653.5 million. Total beer taxes, at $18 per barrel (31 gallons), amounted to $3.37 billion. Brewers who produce less than 2 million barrels get a reduced tax rate ($7 per barrel) on the first 60,000 barrels. (See Table 8.1.)

In 1999 state tax collected on alcoholic beverages averaged about $14.33 per capita. Total state collections from alcohol include not only the tax on beverages ($3.9 billion in 1999) but also payments for alcoholic beverage licenses ($304.7 million). Taxes on alcohol amounted to almost 1 percent of the nation's total state taxes collected in 1999. (See Table 8.2.)

Tobacco Taxes

In 2000, federal, state, and local governments collected $14.5 billion in excise taxes on tobacco products. (See

Table 8.3.) About 97 percent came from cigarette sales. In 2000 the average state tax was 41.9 cents a pack, ranging from 2.5 cents in Virginia, one of the top tobacco-producing states, to $1.11 a pack in New York. In the past seven years, state excise taxes have risen significantly as many states increased taxes. Nineteen states impose taxes of 50 cents per pack or more. Figure 8.1 shows the state tobacco taxes that were in effect as of July 1, 2000.

In 2000 the federal excise tax rose from 24 cents a pack to 34 cents a pack. In 1991, in contrast, the federal excise tax on cigarettes was 16 cents per pack. In May 2001 a presidential commission comprised of farmers, health advocates, and economic experts recommended to President George W. Bush that the federal tax on cigarettes should increase by 17 cents a pack. This increased tax would be used to pay tobacco farmers to stop growing the crop. After five years, the tax would be used for smoking cessation programs and other public health initiatives. As of July 2001, the excise tax is scheduled to rise by five cents in 2002.

GOVERNMENT LEGISLATION AND REGULATIONS

In addition to taxation, the alcoholic beverage and tobacco industries are subject to a vast array of federal and state laws, with regulations covering everything from sales and advertising to shipping.

Alcohol Regulation

The best-known pieces of legislation regarding alcohol are the Eighteenth and Twenty-first Amendments to the Constitution. The Eighteenth Amendment prohibited the manufacture, sale, and importation of alcoholic beverages. Ratified in 1919, it took effect in 1920, ushering in a period in American history known as Prohibition. After 12 years, during which it failed to stop the manufacture and

TABLE 8.1

Alcohol, Tobacco, and Firearms (ATF) tax collections, fourth quarter cumulative summary

Source of Revenue	Rate	Amount (000's)	
		FY 1999	FY 1998
Excise tax, total		$13,180,147	$13,425,611
Alcohol tax, total		$7,712,200	$7,594,320
Distilled Spirits Tax, Total		$3,684,392	$3,539,710
Domestic	$13.50 per proof gallon	$2,925,666	$2,857,430
Imported	$13.50 per proof gallon	$758,726	$682,280
Wine Taxes, Total		$653,518	$634,142
Domestic	Various	$498,642	$480,265
Imported	Various	$154,876	$153,877
Beer Taxes, Total		$3,374,290	$3,420,468
Domestic	$18 or $7 per barrel	$3,052,973	$3,147,823
Imported	$18 per barrel	$321,317	$272,645
Tobacco tax, total		$5,300,499	$5,672,908
Domestic	Various	$5,185,975	$5,608,259
Imported	Various	$114,524	$64,649
Firearms and ammunition tax, total	10% or 11% of sales price	$167,448	$158,383
Special (occupational) tax, total		$104,737	$106,244
Total tax collections		**$13,284,884**	**$13,531,855**

Note: All "imported" figures are obtained from U.S. Customs data. Source for the other figures is a database that records collections by tax return period. The data is summarized on this table by quarter for which liability was incurred.

SOURCE: "Alcohol, Tobacco, and Firearms Tax Collections, Cumulative Summary Fourth Quarter Fiscal Year," Bureau of Alcohol, Tobacco, and Firearms, U.S. Department of the Treasury, Washington, DC, 1999

TABLE 8.2

Tax collections by state governments, 1999

(amounts in thousands of dollars)
(per capita amounts in whole dollars)

Code	Description	Total*	Per Capita	Population
	United States, Total Taxes	499,510,046	1,835.27	272,172
T01	Property Taxes	11,256,732	41.36	
	Sales and Gross Receipts Taxes, Total	239,871,421	881.32	
T09	General Sales and Gross Receipts	165,717,430	608.87	
	Selective Sales Taxes, Total	74,153,991	272.45	
T10	Alcoholic Beverages	3,901,092	14.33	
T11	Amusements	2,824,152	10.38	
T12	Insurance Premiums	9,569,469	35.16	
T13	Motor Fuels	29,200,869	107.29	
T14	Parimutuels	381,390	1.40	
T15	Public Utilities	8,889,053	32.66	
T16	Tobacco Products	8,190,095	30.09	
T19	Other Selective Sales	11,197,871	41.14	
	License Taxes, Total	30,403,564	111.71	
T20	Alcholic Beverage License	304,692	1.12	
T21	Amusement License	282,872	1.04	
T22	Corporation License	6,358,764	23.36	
T23	Hunting & Fishing License	1,075,852	3.95	
T24	Motor Vehicle License	14,074,525	51.71	
T25	Motor Vehicle Operators License	1,265,702	4.65	
T27	Public Utility License	365,788	1.34	
T28	Occupation & Business Licenses, not elsewhere classified	6,229,512	22.89	
T29	Other Licenses	445,857	1.64	
	Other Taxes, Total	217,978,329	800.89	
T40	Individual Income	172,341,998	633.21	
T41	Corporation Net Income	30,692,483	112.77	
T50	Death & Gift	7,493,136	27.53	
T51	Documentary & Stock Transfer	4,089,092	15.02	
T53	Severance	3,127,662	11.49	
T99	All Other	233,958	0.86	

*Includes the 50 State governments only. Does not include the District of Columbia or any local government.

SOURCE: "State Government Tax Collections: 1999," Bureau of the Census, U.S. Department of Commerce, Washington, DC, 2000

TABLE 8.3

Governmental revenues from tobacco products, 1991–2000[1]

In millions of dollars

Year	Excise taxes			Total excise taxes	State sales tax
	Federal	State	Local		
1991	5,062	6,130	198	10,972	1,469
1992	5,185	6,200	194	11,436	1,996
1993	5,563	6,472	188	12,062	2,042
1994	5,977	7,025	185	12,623	2,005
1995	5,860	7,535	182	13,342	2,000
1996	5,913	7,636	181	13,730	2,013
1997	5,839	7,750	177	13,766	2,003
1998	5,475	7,975	196	13,646	2,181
1999	5,306	7,962	195	13,463	2,411
2000	5,973	8,357	195	14,525	2,371

[1]Calendar year. Includes imports.

SOURCE: "Table 31: Governmental revenues from tobacco products, 1991–2000," in *Tobacco Situation and Outlook Report*, Economic Research Service, U.S. Department of Agriculture, Washington, DC, April 2001

sale of alcohol, Prohibition was repealed in 1933 by the Twenty-first Amendment.

Most interpretations of the Twenty-first Amendment hold that the amendment gives individual states the power to regulate and control alcoholic beverages within their own borders. Consequently, every state currently has its own alcohol administration and enforcement agency. "Control states" directly control the sale and distribution of alcoholic beverages within their borders. For example, Pennsylvania and Utah act as both wholesalers and retailers of alcoholic beverages through the operation of state-owned package stores. Some critics of this policy have questioned whether such state monopolies violate antitrust laws. Thirty-three states allow only licensed businesses to operate as wholesalers and retailers. Some states have separate administrative agencies for alcohol law enforcement and for collection of alcohol tax revenues.

DIRECT SHIPMENTS—RECIPROCITY OR FELONY? A legislative controversy has developed over the direct shipment of alcoholic beverages from one state to consumers or retailers in another. The Twenty-first Amendment states that "the transportation or importation into any state, territory, or possession of the United States for delivery or use therein of intoxicating liquors, in violation of the laws thereof, is hereby prohibited." Because the laws of the states are not uniform, several states passed reciprocity legislation, allowing specific states to exchange direct shipments, thus eliminating the state-licensed wholesalers from the exchange.

Wholesalers have charged that reciprocity and direct shipment are violations of the Twenty-first Amendment. Several state legislatures, including those of Florida, Georgia, and Kentucky, passed laws repealing their state reciprocity agreements and making direct shipment a felony. As of July 2001, 14 other states were considering

such legislation. Only 13 states have reciprocity laws allowing residents to "import" alcoholic beverages for their own consumption from other states. The other 37 states forbid it.

Increases in sales of alcoholic beverages have resulted from Internet sales and from catalog wine clubs and beer-of-the-month clubs. In 1999 a preliminary version of the "21st Amendment Enforcement Act" passed in both the House of Representatives and the Senate. This legislation would make it difficult for companies to sell alcohol over the Internet or through mail-order services. It would allow state attorneys general in states that ban direct alcohol sales to seek a federal injunction against companies that violate their liquor sales laws. The bill awaits a House-Senate conference to shape its final version.

Within a month of the passage of this legislation, the high-tech community voiced its concern over such legislation, suggesting that if states could ban Internet wine sales, they might restrict other electronic commerce as well. Senator Orrin Hatch (R-Utah) said he crafted the bill to take other e-commerce concerns into account, and insisted that the measure is "narrowly tailored" to deal with alcohol only.

Wholesalers warn that teenagers can purchase alcohol online by using their parents' credit cards. Wineries respond that minors do not drink wine, the alcoholic beverage most often offered online, and that they want alcohol right away, not in the week or two it would take to receive that delivery. They believe the wholesalers support the proposed legislation because they want to continue to maintain control of alcohol distribution in their states.

Tobacco Regulation

Federal tobacco legislation has covered everything, from unproved advertising claims and warning label

FIGURE 8.1

Cigarette excise taxes, July 1, 2000

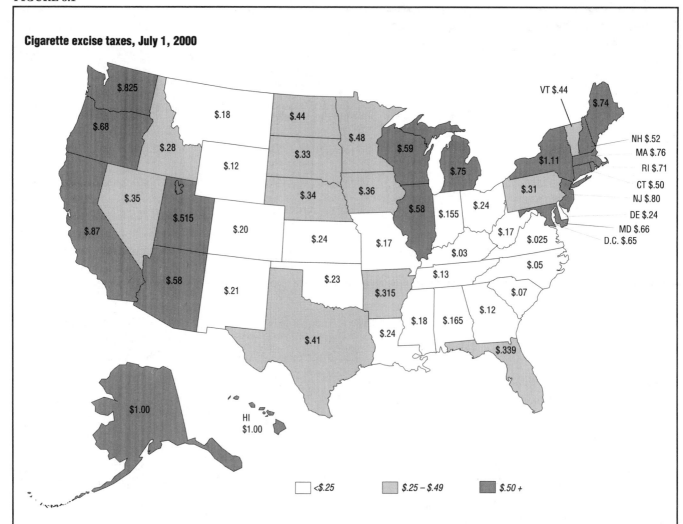

Rank	State	Cents/pack
1	New York	111.0
2	Alaska	100.0
2	Hawaii	100.0
4	California	87.0
5	Washington	82.5
6	New Jersey	80.0
7	Massachusetts	76.0
8	Michigan	75.0
9	Maine	74.0
10	Rhode Island	71.0
11	Oregon	68.0
12	Maryland	66.0
13	District of Columbia	65.0
14	Wisconsin	59.0
15	Arizona	58.0
15	Illinois	58.0
17	New Hampshire	52.0
18	Utah	51.5
19	Connecticut	50.0
20	Minnesota	48.0
21	North Dakota	44.0
21	Vermont	44.0
23	Texas	41.0
24	Iowa	36.0
25	Nevada	35.0
26	Nebraska	34.0

Rank	State	Cents/pack
27	Florida	33.9
28	South Dakota	33.0
29	Arkansas	31.5
30	Pennsylvania	31.0
31	Idaho	28.0
32	Delaware	24.0
32	Kansas	24.0
32	Louisiana	24.0
32	Ohio	24.0
36	Oklahoma	23.0
37	New Mexico	21.0
38	Colorado	20.0
39	Mississippi	18.0
39	Montana	18.0
41	Missouri	17.0
41	West Virginia	17.0
43	Alabama	16.5
44	Indiana	15.5
45	Tennessee	13.0
46	Wyoming	12.0
46	Georgia	12.0
48	South Carolina	7.0
49	North Carolina	5.0
50	Kentucky	3.0
51	Virginia	2.5
	Average State Tax	**41.9**

SOURCE: "Cigarette Excise Taxes, July 1, 2000," in *Investment in Tobacco Control: State Highlights–2000*, U.S. Department of Health and Human Services, Centers for Disease Control and Prevention, and National Center for Chronic Disease Prevention and Health Promotion, Office on Smoking and Health, Atlanta, GA, 2001

requirements, to the development of cigarettes and little cigars that are less likely to start fires. In the past, the Food and Drug Administration (FDA) has prohibited the claim that Fairfax cigarettes prevented respiratory and other diseases (1953), and denied the claim that tartaric acid, which was added to Trim Reducing-Aid cigarettes, helped to promote weight loss (1959).

The Federal Trade Commission (FTC) has also been given jurisdiction over tobacco issues in several areas. As early as 1942, the FTC had issued a "cease-and-desist" order in reference to Kool cigarettes' claim that smoking Kools gave extra protection against or cured colds. In January 1964 the FTC proposed a rule to strictly regulate cigarette advertisements and to prohibit explicit or implicit health claims by cigarette companies.

On the other hand, for many years, federal legislation avoided restrictive orders concerning the use of tobacco products. For example, the terms "consumer product" and "hazardous substance" were specifically defined to exclude tobacco products from regulation by several legislative acts during the 1970s and 1980s. By the mid-1980s, however, legislators were considering tobacco more critically.

Some of the legislation of the late 1980s included requiring four alternating health warnings to be printed on tobacco packaging, prohibiting smokeless tobacco advertising on television and radio, and banning smoking on domestic airline flights. In 1993 the Environmental Protection Agency (EPA) released its final risk assessment on environmental tobacco smoke (ETS, or "secondhand smoke") and classified it as a known human carcinogen (cancer-causing agent). In 1994 the Occupational Safety and Health Administration (OSHA) proposed regulations that would prohibit smoking in workplaces, except in smoking rooms that are separately ventilated.

FDA REGULATION OF TOBACCO PRODUCTS. The Food and Drug Administration (FDA) investigated the tobacco industry to determine whether nicotine is an addictive drug that should be regulated like other addictive drugs. Weeks of testimony before Congress indicated that tobacco companies may have been aware of the addictive effects of nicotine and the likely connection between smoking and cancer as early as the mid-1950s.

At first, representatives from the tobacco companies and the Tobacco Institute (which represented the major tobacco companies) denied that there was any scientifically proven connection between smoking and cancer and claimed that nicotine was not addictive. The industry acknowledged that it changed levels of nicotine in cigarettes, but for taste purposes only, not to increase the addictiveness of the products.

In August 1995 the FDA ruled that the nicotine in tobacco products is a drug and, therefore, liable to FDA

regulation. Based on this finding, the Clinton administration proposed regulations designed to stop the sale of cigarettes to minors. President Clinton recommended that cigarette sales to anyone under 18 be prohibited, that the sale of cigarettes through vending machines be banned, and that advertising in media designed for minors be stopped.

Some of the provisions of the proposed FDA regulations were

- The minimum age for purchasing tobacco products would be 18.

- Vending machines and self-service displays would be banned, except in nightclubs and other facilities that bar the admission of persons under age 18.

- "Kiddie packs" (containing five or fewer cigarettes) and free samples would be banned.

- Advertising billboards would have to be one thousand or more feet from schools and playgrounds.

- Other billboards, in-store advertising, and outdoor displays would be limited to black-and- white text only— no colors or graphics, except in "adult-only" facilities where the advertising cannot be seen from the outside or removed from its display location.

- Brand-name sponsorship of sporting or other public events would be prohibited; only company-name sponsorship would be allowed. Also, it would be prohibited to print brand names on hats, T-shirts, and other personal items.

The FDA recommendations were to go into effect on August 28, 1996, but a lawsuit by the tobacco, advertising, and convenience store industries against the FDA delayed the implementation of the FDA order. In April 1997 the Federal District Court in Greensboro, North Carolina, in *Brown and Williamson Tobacco Corp., et al. v. Food and Drug Administration, et al.* (966 F Supp 1374), ruled that the FDA did have jurisdiction under the Federal Food, Drug, and Cosmetic Act (52 Stat 1040) to regulate cigarettes and smokeless tobacco products, upholding the restrictions that involved youth access and labeling. However, the court found that the FDA did not have authority to regulate the advertising of tobacco products. Both parties appealed the district court decision.

In August 1998 the U.S. Court of Appeals for the Fourth Circuit in Richmond, Virginia, reversed the district court's decision. In *Brown and Williamson Tobacco Corp., et al. v. Food and Drug Administration, et al.* (153 F.3d 155), the court ruled that the "FDA lacks jurisdiction to regulate tobacco products" and that "all of the FDA's August 28, 1996, regulations of tobacco products are thus invalid."

On April 26, 1999, the U.S. Supreme Court granted the FDA's Petition for a Writ of Certiorari to review the decision of the U.S. Court of Appeals for the Fourth

Circuit (67 LW 3652). The granting of the petition allowed the age and identification provision of FDA's tobacco regulations to remain in effect pending the Court's final decision. On March 21, 2000, the Supreme Court ruled (5-4) that the government lacks authority to regulate tobacco as an addictive drug. Although the FDA cannot regulate tobacco, state laws on selling cigarettes to minors are not affected by the ruling.

SHOULD THE U.S. MILITARY SUBSIDIZE CIGARETTE SALES. In 1995, according to a report prepared by the U.S. Department of Defense (DoD), the Pentagon spent about $1 billion of taxpayer money for health and work expenses related to tobacco products. These costs included

- $345 million in lost work productivity due to cigarette breaks.

- $1 million for active-duty hospitalization for smoking-related conditions.

- $584 million for smoking-related health care for military beneficiaries aged 35 to 85.

The report stated that smoking is encouraged by the below-market prices charged for cigarettes at the 230 military base commissaries (grocery stores). Commissaries sell about 58 million cartons of cigarettes a year, at prices 30 to 60 percent below the prices charged in civilian grocery stores. In 1995 at the Fort McCoy (Wisconsin) commissary, cigarettes accounted for half of total sales, while nearby commercial grocers reported that cigarettes accounted for only 3 percent of sales.

In response to the situation, the inspector general of the DoD, Eleanor Hill, announced that prices at commissaries would go up by $4.00 a carton. The tobacco industry, fearing an estimated $200 million in lost sales to the commissaries, immediately lobbied Congress to stop the price increase. Nevertheless, the policy went into effect on November 1, 1996.

STATE REGULATIONS. In 1998 the Centers for Disease Control and Prevention (CDC) and the National Cancer Institute reviewed state tobacco laws regulating smoke-free indoor air, youth access to tobacco products, advertising, and excise taxes. The survey identified state laws related to tobacco control in effect as of December 31, 1998. Forty-six states and Washington, D.C., restricted environmental tobacco smoke to some degree or in certain places, all states taxed cigarettes and prohibited the sale of tobacco products to minors, and thirteen states restricted tobacco advertising.

In 2001 the American Lung Association developed comparison maps that show the restrictions on smoking in public places in 1991 as compared to 2000. As Figure 8.2 and Figure 8.3 show, seven states had no restrictions on smoking in public places in 1991. By 2000, only one state

remained that posed no restrictions. At the other end of the restriction scale, no states had comprehensive smoking restrictions in 1991 and only one state had extensive restrictions, but by 2000, four states had enacted comprehensive restrictions and eight states extensive restrictions.

Other Government Programs

Antismoking campaigns, including those launched by the U.S. surgeon general, are taking their toll on the tobacco industry, particularly on farmers for whom tobacco is their primary source of income. In an effort to prevent the complete collapse of these farms, the government has helped many farmers diversify their crops. However, agricultural specialists are skeptical that enough of a market exists for a major changeover to other crops. Even if it did, few crops can match the financial return per acre of tobacco.

The U.S. Department of Agriculture (USDA) has continued its efforts to stabilize tobacco production by administering marketing quotas to limit the amount of tobacco produced, thus artificially maintaining higher prices.

SUING THE TOBACCO COMPANIES

Between 1960 and 1988, close to 300 lawsuits sought damages from tobacco companies for smoking-related illnesses; however, courts consistently held that people who choose to smoke are responsible for the health consequences of that decision. But in 1988, for the first time, a tobacco company was ordered to pay damages. A federal jury in Newark, New Jersey, ordered Liggett Group, Inc., to pay $400,000 to the family of Rose Cipollone, a longtime smoker who died of lung cancer in 1984. The case was overturned on appeal, but the Supreme Court ruled in favor of the Cipollone family in *Cipollone v. Liggett Group, Inc.* (505 U.S. 504, 1992). In the 7-2 ruling, the Court broadened a smoker's right to sue cigarette makers in cancer cases. The justices decided that the Federal Cigarette Labeling and Advertising Act of 1966 (PL 89-92), which required warnings on tobacco products, did not preempt damage suits. Despite the warnings on tobacco packaging, people can still sue on the grounds that tobacco companies purposely concealed information about the risks of smoking.

The "Tobacco Wars"

Since 1992, individuals, states, and cities have filed numerous lawsuits against the nation's tobacco companies. Individuals and their families have sued because of the damage caused to individuals by their smoking and by their inhalation of environmental tobacco smoke (ETS). About 40 states, the cities of New York and San Francisco, the Commonwealth of Puerto Rico, and other third parties involved in paying for medical care have sued to try to recover Medicaid, Medicare, and other medical costs caused by tobacco-related health problems.

FIGURE 8.2

Restrictions on smoking in public places, 1991

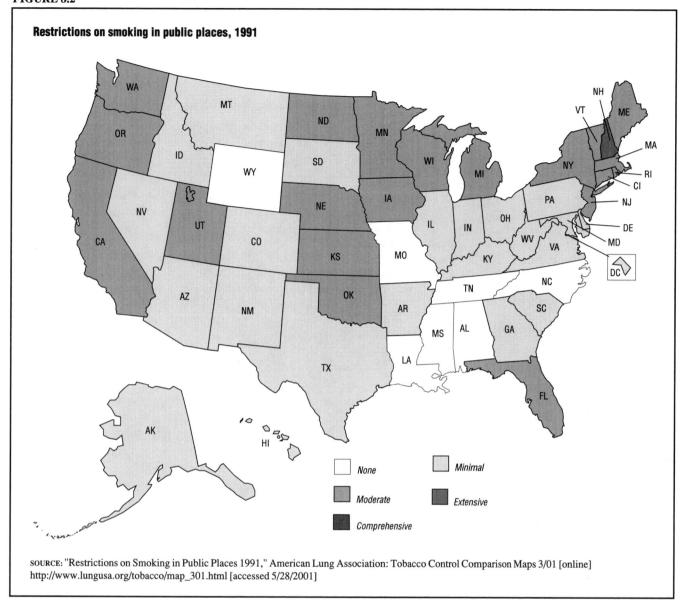

SOURCE: "Restrictions on Smoking in Public Places 1991," American Lung Association: Tobacco Control Comparison Maps 3/01 [online] http://www.lungusa.org/tobacco/map_301.html [accessed 5/28/2001]

Jurors at first found it difficult to side with individuals who had knowingly used a product linked to so many health problems. But juries found more merit in individuals' suits after the discovery of documents that implied tobacco companies knew about the addictiveness and health hazards of tobacco products for many years, and that they had changed these products to enhance their addictiveness.

A POTENTIAL SETTLEMENT. To avoid the onslaught of lawsuits, the tobacco industry sought a national settlement with the states, in return for future protection from lawsuits. In June 1997 the country's largest tobacco companies and 40 states that had filed suit against the tobacco industry agreed on a settlement. According to the proposed agreement, tobacco companies would pay the states $368.5 billion over 25 years to reimburse them for their tobacco-related medical costs and to pay for tobacco-control programs to reduce tobacco use among teenagers.

In addition, tobacco companies would accept FDA authority to regulate tobacco products, restrict their advertising, and release internal research documents related to the health effects of their products. The states would drop all claims against the tobacco companies and grant the industry immunity from future class-action lawsuits (suits on behalf of large groups of people). This proposed settlement required changes in federal law before taking effect.

Over the next several months, various members of Congress introduced their own comprehensive tobacco bills, broadly based on the settlement. All failed, and the states resumed negotiation with the tobacco companies to try to reach a more limited settlement, one that would not require federal legislation to take effect.

MORE STATE LAWSUITS. In the meantime, the industry settled two of its pending state lawsuits. In the summer of 1997, tobacco companies agreed to pay Mississippi

FIGURE 8.3

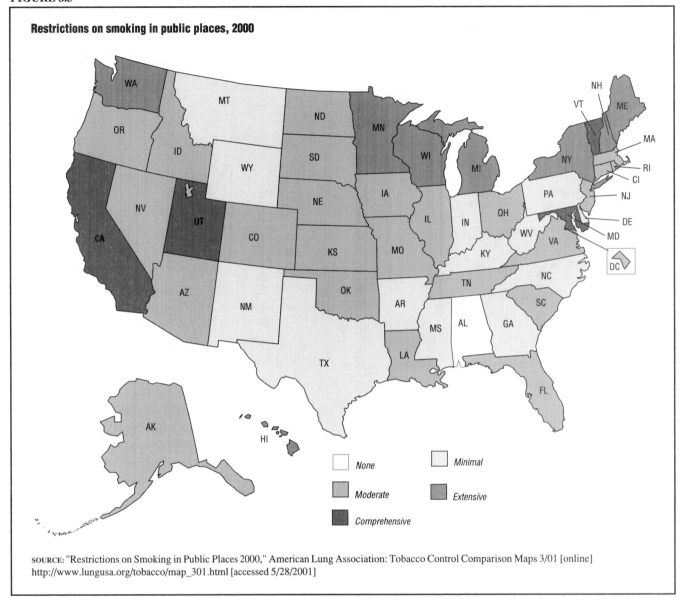

Restrictions on smoking in public places, 2000

Legend:
- None
- Minimal
- Moderate
- Extensive
- Comprehensive

SOURCE: "Restrictions on Smoking in Public Places 2000," American Lung Association: Tobacco Control Comparison Maps 3/01 [online] http://www.lungusa.org/tobacco/map_301.html [accessed 5/28/2001]

$3.4 billion and Florida $11.3 billion. In the Florida settlement, they agreed to remove tobacco billboards, public transit advertising, and vending machines, and to release internal documents.

The tobacco companies had long denied that nicotine was addictive, that they had manipulated nicotine levels to make their products more addictive, and that they had targeted young people in their advertising campaigns. They had also denied that scientific research had demonstrated links between health problems and tobacco products. In August 1997, however, two important events occurred in pretrial testimony for the lawsuit brought by the state of Florida.

First, the chairman of the Philip Morris Companies, Geoffrey Bible, testified that smoking-related diseases "might have" caused the deaths of 100,000 Americans in recent decades. Second, Steven F. Goldstone, chairman of

RJR Nabisco Holding Corporation (the parent company of the R.J. Reynolds Tobacco Company), testified that he believed cigarettes "play a role in causing lung cancer." Not only did the cigarette manufacturers know that cigarettes can cause severe and sometimes fatal health problems, but later releases of tobacco industry documents showed that they actually enhanced nicotine delivery in some brands to make them more competitive. Indicative of the industry's concessions that it had released misleading information, the Tobacco Institute, which disseminated information for the tobacco industry, was shut down in January 1999.

In early 1998 the industry settled with Texas for $15.3 billion. In May, after having gone to trial, but just before the case was to go to the jury, the tobacco companies settled with Minnesota for $6.6 billion. The Minnesota case forced the disclosure of millions of internal tobacco company documents, exposing deceptive conduct and laying the foundation for future legal actions. The Minnesota agreement required the industry to maintain depositories of these documents and to release an index to millions of previously released documents.

THE TOBACCO MASTER SETTLEMENT AGREEMENT

On November 23, 1998, attorneys general from 46 states (excluding the four states that had previously settled), five territories, and the District of Columbia signed an agreement with the major tobacco companies to settle all the state lawsuits to recover the Medicaid costs of treating smokers. In addition to restrictions on tobacco advertising, marketing, and promotion, the Master Settlement Agreement (MSA) requires the tobacco companies to make annual payments totaling about $200 billion over the next 25 years, beginning in the year 2000.

In a side agreement to the national tobacco settlement, a $5.15 billion trust was established between tobacco-growing states and major cigarette makers. This trust creates a pool of money separate from the Master Settlement Agreement to help tobacco farmers who will lose money as a result of the settlement.

The MSA is much more limited in scope than the June 1997 proposed settlement. Table 8.4 compares the Master Settlement Agreement with the earlier proposal. There are several important differences between the two documents. For one thing, the MSA is a private contract between the industry and the states and does not require federal legislation. Instead, the National Association of Attorneys General (NAAG) will manage the MSA for the states.

In addition, the MSA settles only the state and local government lawsuits. The tobacco industry is still liable for class action and individual lawsuits. Furthermore, the MSA lacks many of the earlier proposal's tobacco-control initiatives, such as restrictions on advertising, marketing, and promotion, and does not allow for FDA regulation of tobacco products.

Supporters of the tobacco agreement maintain that the settlement will have a significant impact on reducing underage tobacco use. Critics assert that the MSA accomplished little other than a transfer of wealth from smokers (through the tobacco companies) to state treasuries. The major cigarette manufacturers raised prices by 45 cents a pack to cover the cost of the annual payments required by the settlement.

Receiving the Money

States must gain State Specific Finality and enact the Model Statute before receiving all of their expected payments from the participating tobacco manufacturers (Brown and Williamson, Lorillard, Philip Morris, R.J. Reynolds, Commonwealth Tobacco, and Liggett and Myers). State Specific Finality is achieved when a state court approves the Master Settlement Agreement and the required consent decree containing many of the provisions of the agreement. This must be final approval, with no appeals pending and all outstanding suits against the tobacco industry settled. Funds will become available when 80 percent of the states achieve State Specific Finality.

The Model Statute, to be enacted by the state legislatures, assesses a per-pack fee on nonparticipating cigarette manufacturers to be put into a reserve fund. Such a requirement protects the market share of the participating companies. If a state fails to enact such a statute, reductions in that state's allocation of monies of up to 65 percent will result.

The settlement funds are allocated to the recipients according to a formula developed by the state attorneys general. The formula is based on estimated tobacco-related Medicaid expenditures and the number of smokers in each state. Table 8.5 shows the total amounts to be paid to each state through 2025. Seven states—Arizona, Arkansas, California, New Jersey, New York, Pennsylvania, and Tennessee—have legal challenges to the settlement. The remaining 39 states have received final court approval or, facing no legal challenges, are waiting for the appeals deadline to pass.

The funds are subject to a number of adjustments, reductions, and offsets, such as the volume-of-sales adjustment. If, as anticipated by public health officials, cigarette sales decline as a result of higher prices, the annual payments will be reduced proportionately.

States had been concerned that the federal government would attempt to recoup some of the states' tobacco settlement revenues. However, the Fiscal Year 1999 Emergency Supplemental Appropriations Bill (PL 105-277) included an amendment prohibiting the federal government from taking any of the states' tobacco settlement money.

Spending the Money

The Master Settlement Agreement does not tell states how to spend the money they receive from the tobacco companies. Nearly 500 tobacco-settlement bills have been introduced in state legislatures. Most deal with establishing a trust fund to allocate money to specific issues, such as tobacco control and smoking cessation programs, children's health, education, and highway construction. Other bills propose using the settlement funds to compensate tobacco farmers or to pay for long-term care or tax cuts. Anti-smoking advocates are critical of states' plans to use the money for programs unrelated to tobacco control.

Issues for Congressional Consideration

Several issues were unresolved by the MSA. It is likely that tobacco legislation will continue to be introduced

TABLE 8.4

Comparison of Master Settlement Agreement (MSA) with the June 1997 Proposal

Topic	MSA	June 1997 Proposed Settlement
Advertising, Marketing, and Promotion	Prohibits targeting youth. Bans use of cartoons. Permits corporate sponsorship of sporting and cultural events. Limits companies to one brand-name sponsorship a year (may not include team sports, events with a significant youth audience, or events with underage contestants). Bans public transit advertising. Bans billboard advertising in arenas, stadiums, malls, and arcades. Allows billboard advertising for brand-name sponsored events. Limits advertising outside retail stores to signs no bigger than 14 sq. ft. Bans payments to promote tobacco products in various media. Bans non-tobacco merchandise with brand-name logos except at brand-name sponsored events. Bans gifts of non-tobacco items to youth in exchange for tobacco products. Restricts use of non-tobacco brand names for tobacco products.	Includes the following additional prohibitions and restrictions: Bans use of cartoon and human images. Bans all outdoor advertising and brand-name sponsorship. Bans Internet advertising. Restricts point-of-sale advertising. Restricts permissible tobacco advertising to black text on a white background except in adult-only facilities and adult publications. Bans non-tobacco merchandise with brand-name logos. Bans non-tobacco items, gifts, and services. Restricts self-service displays to adult-only facilities. Requires all tobacco products be placed out of reach of customers except in adult-only facilities.
Youth Access	Limits free samples to adult-only facilities. Bans sale of cigarettes in packs of less than 20 through December 2001.	Bans free samples and vending machines. Sets minimum age of 18 (verified with photo ID) to purchase tobacco. Face-to-face sales only. Mandates minimum pack size of 20 cigarettes.
Corporate Culture	Requires corporate commitments to reducing youth access and consumption. Prohibits manufacturers from suppressing health research. Disbands existing tobacco trade associations and provides regulation and oversight of new trade organizations.	Requires corporate commitments to reducing youth access and consumption. Protects industry whistle-blowers. Disbands existing tobacco trade associations. Prohibits manufacturers from suppressing health research.
Industry Lobbying Restrictions	Companies agree not to lobby against certain specified kinds of state anti-tobacco legislation and regulation, but permits them to oppose efforts to raise excise taxes, create lookback penalties, or restrict ETS exposure. Requires lobbyists to seek company authorization for their activities.	Requires lobbyists to seek company authorization for their activities.
Tobacco Document Disclosure	Industry agrees to release, and create a website for, all documents under protective orders in specified state lawsuits, except those for which companies assert privilege or trade-secret protection.	Establishes a public depository of industry documents. Industry must provide a detailed log of documents determined to be privileged against disclosure. Establishes an arbitration panel to settle disputes over making privileged documents public.
Annual Payments	Mandates up-front and annual payments to the states totaling $204.5 billion through 2025. Payments subject to inflation adjustment, volume-of-sales adjustment, and a federal legislation adjustment. No restrictions on how the states spend the funds.	Mandates up-front and annual payments totaling $368.5 billion over the first 25 years, allocated as follows: $193.5 billion in unrestricted funds to states; $36 billion for cessation; $25 billion for research; $37 billion for tobacco control; $77 billion (if required) to settle lawsuits. Payments subject to inflation and volume-of-sales adjustment.
Anti-Tobacco Research and Education	Creates a national foundation to reduce underage tobacco use and substance abuse. Requires industry to pay the foundation $250 million over 10 years to fund research and surveillance, and $1.45 billion (subject to inflation and volume-of-sales adjustment) over 5 years to pay for a national anti-tobacco education program.	Provides $25 billion over 8 years (see above) to create a public health trust to fund biomedical and behavioral tobacco-related research. Provides $37 billion (see above) over first 25 years to fund tobacco-control programs, including anti-tobacco advertising ($0.5 billion/yr), FDA regulation, and local community activities.
Enforcement, Consent Decrees	Requires companies and states to sign legally enforceable consent decrees that include key provisions of the agreement. Only the tobacco divisions and not the parent companies are liable. Mandates the National Association of Attorneys General to coordinate implementation and enforcement of the agreement. Directs industry to pay $52 million for that purpose.	Requires companies and states to sign legally enforceable consent decrees that include key provisions of the agreement. Establishes civil and criminal penalties for violations of the agreement.
Attorneys' Fees	Companies agree to pay all fees and expenses of attorneys general, subject to a $150 million annual cap. Requires companies to pay outside attorneys retained by states: either (i) all fees paid from a $1.25 billion pool, or (ii) fees determined by arbitration panel and paid subject to a $500 annual cap.	Requires companies to pay all fees and expenses of outside attorneys retained by states. Establishes an arbitration panel to determine and award attorneys' fees and expenses.
Civil Liability	Settles state and local medical-cost reimbursement lawsuits and protects industry (including retailers and distributors) from future state and local tobacco-related lawsuits. Allows dollar-for-dollar reduction in state's recoveries should the industry be found liable in a local government lawsuit.	Settles state medical-cost reimbursement and class-action lawsuits. Prohibits future class actions. Prohibits punitive damage awards in individual lawsuits arising from past industry conduct. Caps total annual liability at $5 billion.

SOURCE: "Table 8.6: Comparison of Master Settlement Agreement (MSA) with the June 1997 Proposal," in *Tobacco Master Settlement Agreement (1998): Overview and Issues for the 106th Congress*, Congressional Research Service, The Library of Congress, Washington, DC, 1999

in Congress. Public health officials and others want a "youth look-back" provision that would keep track of tobacco use by minors and penalize the industry if use does not decline. Anti-smoking advocates will likely continue to push for a tax increase on tobacco products. They maintain that sharp price increases result in a decline in cigarette sales.

INDIVIDUAL AND CLASS-ACTION LAWSUITS

The tobacco industry faces hundreds of individual cases across the country, including cases filed by health insurers. In March 1999 a state jury in Portland, Oregon, in *Branch-Williams v. Philip Morris* (No. 9705-03957, Circuit Court for the County of Multnomah, Portland, Oregon, 1999), ordered Philip Morris to pay $81 million to the family of a man who smoked Marlboro cigarettes for 40 years prior to his death.

A month earlier, in *Henley v. Philip Morris Inc., et al.* (No. 995172, Sup. Ct. of California, 1999), a San Francisco jury awarded $51 million in a case brought by a woman with inoperable lung cancer. Philip Morris filed post-trial motions for a new trial, but was denied.

Nationwide, prior to the Boeken award, juries had awarded damages to individual smokers only six times. Three verdicts were overturned, Branch-Williams and Henley are still on appeal, and one verdict was returned in May 2001 with a then-record $1.72 million in compensatory damages. (Compensatory damages are meant to compensate the party that wins the judgment, to make up for their direct or actual losses. Punitive damages are meant to punish the party that loses the judgment.)

Secondhand Smoke Lawsuits

Some lawsuits have been brought against tobacco companies, claiming that environmental tobacco smoke (ETS), or secondhand smoke, has caused harm to individuals. A Mississippi jury did not find the tobacco industry liable for the cancer that killed a barber, despite claims that the disease was caused by decades of ETS inhalation from his customers. In *Estate of Burl Butler, et al. v. Philip Morris, Inc., et al.* (No. 94-5-53, Jones County, Mississippi Circuit Court, Second Judicial District, 1999), the tobacco lawyers successfully argued that Butler had a family history of cancer and had been exposed to asbestos. They suggested that the cancer might have been caused by factors other than the secondhand smoke.

In *Norma R. Broin, et al. v. Philip Morris, et al.* (No. 91- 49738, Dade County, Florida, Eleventh Judicial Circuit, 1997), the lead plaintiff (the party that brought the lawsuit), Norma Broin, was a nonsmoking former American Airlines flight attendant whose lung cancer was diagnosed in 1989. One of the lawyers, Stanley Rosenblatt, claimed flight attendants are "totally innocent victims.

TABLE 8.5

Aggregate Master Settlement Agreement (MSA) payments to the states and territories through 2025

	Dollars in millions		Dollars in millions
Alabama	3,166	New Hampshire	1,305
Alaska	669	**New Jersey**	7,576
Arizonaª	2,888	New Mexico	1,168
Arkansas	1,622	**New York**	25,003
California	25,007	North Carolina	4,569
Colorado	2,686	North Dakota	717
Connecticut	3,637	Ohio	9,869
Delaware	775	Oklahoma	2,030
DC	1,189	Oregon	2,248
Georgia	4,809	**Pennsylvania**	11,259
Hawaii	1,179	Rhode Island	1,408
Idaho	712	South Carolina	2,305
Illinois	9,119	South Dakota	684
Indiana	3,996	**Tennessee**	4,782
Iowa	1,704	Utah	872
Kansas	1,633	Vermont	806
Kentucky	3,450	Virginia	4,006
Louisiana	4,419	Washington	4,023
Maine	1,507	West Virginia	1,737
Maryland	4,429	Wisconsin	4,060
Massachusetts	7,913	Wyoming	487
Michigan	8,526	American Samoa	30
Missouri	4,456	Guam	43
Montana	832	N. Marianas	17
Nebraska	1,166	U.S. Virgin Islands	34
Nevada	1,195	Puerto Rico	2,197
Total			**195,919**

NOTE: The table does not include the 4 states that settled individually with the tobacco companies (i.e., Florida, Minnesota, Mississippi, Texas). Allocation of payments is based on each state's Medicaid expenditures and the number of smokers in each state. The 7 states listed in boldface have legal challenges to the settlement. The remaining 39 states have either received final court approval or face no legal challenges and are waiting for the appeals deadline to pass.
ªPayments include the up-front payment (paid in 5 installments through 2003) and the annual payments through 2025. The annual payments do not reflect any of the adjustments and reductions set out in the MSA (e.g., inflation adjustment, volume-of-sales adjustment, federal legislation adjustment). An additional $8.61 billion is to be paid in 10 annual installments (2008-2017) to the strategic contribution fund and allocated to states to reflect their contribution toward resolution of the state lawsuits against the tobacco companies.

SOURCE: "Table 8.7: Aggregate MSA Payments ($ in millions) to the States and Territories through 2025*," in *Tobacco Master Settlement Agreement (1998): Overview and Issues for the 106th Congress*, Congressional Research Service, The Library of Congress, Washington, DC, 1999

They are forced to involuntarily suck in the smoke of other people while the tobacco industry has lied for years about the dangers." The 60,000 flight attendants represented by Broin claimed injury from breathing secondhand smoke before the 1990 ban on in-flight smoking. In October 1997 the lawsuit was settled mid-trial for $349 million. The money paid in damages will be used for research into diseases associated with tobacco smoke, and their early detection and treatment.

First Class-Action Lawsuit on Behalf of Smokers

A class action lawsuit is one brought forward by a group of parties with similar claims. *Engle, et al. v. R. J. Reynolds Tobacco, et al.* (No. 94-08273, Dade County, Florida, Eleventh Judicial Circuit, 1998) was the first class-action lawsuit to go to trial brought on behalf of smokers. In July 1999, following nearly a year of proceedings, the jury gave a detailed verdict. It found that smoking caused many diseases, including cancer and lung and heart diseases, and that the tobacco industry had committed fraud, misrepresentation, conspiracy, negligence, and intentional infliction of emotional distress—and therefore, was liable for punitive damages.

The second phase of the trial (Phase II), to establish the punitive (punishment) damages as well as individual damages for the nine named plaintiffs, concluded with verdicts in favor of the class in 1999. The first stage of Phase II, the trial on compensatory damages for certain class representatives, ended with verdicts in favor of three of the named plaintiffs. The second stage of Phase II, the trial on the amount of punitive damages, concluded with verdicts totaling $145 billion. R.J. Reynolds is appealing this decision and, in May 2001, reached an agreement to pay Florida smokers $709 million regardless of the outcome of the appeal. In exchange, the defendants received a guarantee that the appeals process will be completed without their having to pay the punitive judgment damages during that process.

CHAPTER 9
CAFFEINE

As Americans sit down to their morning cups of coffee or tea or drink a cola, few think about the fact that they are taking a drug. A drug, however, is defined as any substance that affects the mood or the state and function of the body. And according to this definition, the caffeine in many of our favorite beverages is a drug.

SOURCES OF CAFFEINE

Caffeine occurs naturally in the seeds, leaves, or fruits of more than 63 plant species throughout the world. The primary plant sources today, however, are coffee, tea, cacao (cocoa), and kola (cola) plants. These are grown and produced in tropical locations and shipped all over the world. Two other plants, maté and guarana, are primarily grown and consumed in South America as yerba maté and guarana bars and drinks (soda drinks or brewed hot drinks).

Table 9.1 lists the four major sources of caffeine, what part of the plant is consumed, and where the plants are grown. Of the three species of coffee, *coffea arabica L.* accounts for about three-quarters of all coffee consumption. Spices and other substances, such as vanilla and chocolate, can be added to make the currently popular flavored coffees.

The three major tea types (green, oolong, and black) all come from the same species of plant; differences in flavor and color are a result of how the tea leaves are processed. For green tea, the leaves are steamed and dried. For oolong tea, the steamed leaves are partly fermented before they are dried. And for black tea, the steamed leaves are completely fermented before they are dried. As with coffee, spices and other flavorings can also be added to vary the taste of tea.

There are many products on the market today that are called "teas," although they do not contain tea leaves. Virtually any beverage prepared by steeping substances in hot or boiling water is popularly called a "tea." Herbal

TABLE 9.1

Where caffeine comes from

Plant	Species Name	Part of Plant	Major Cultivation Locations
Coffee	*Coffea arabiaca L.* *Coffea robusta* *Coffea liberica*	seeds (beans)	Brazil, Columbia, Indonesia, Ethiopia, Kenya, Jamaica
Tea	*Camellia sinensis*	leaves, buds	India, China
Cacao (cocoa)	*Theobroma cacao*	seeds (beans)	West Africa, Brazil
Kola	*Cola acuminata S.* *Cola nitida*	seeds (nuts)	West Africa

SOURCE: Prepared by the staff of Information Plus; data from several sources.

teas do not contain tea leaves or caffeine; they can make a pleasant substitute for caffeinated drinks.

Varying Amounts of Caffeine

Not all plants contain equal concentrations of caffeine. Different methods of processing and brewing can also affect the amount of caffeine per cup or glass of beverage. When calculating the amount of caffeine per cup or glass, a person must keep in mind the size of the cup or glass and whether he or she dilutes coffee or tea, for example, with milk.

Coffee contains the highest concentration of caffeine. The way it is prepared, however, can change the amount of caffeine it contains. In general, coffee made in a drip-style coffee maker is highest in caffeine content. Tea has significantly less caffeine than coffee, but the longer it is allowed to steep, the higher the caffeine concentration. (See Table 9.2.) Both cocoa and cola drinks generally have lower levels of caffeine ounce per ounce than tea and coffee (see Table 9.2). However, cocoa contains large amounts of theobromine, a chemical related to caffeine that has the same stimulating effect.

TABLE 9.2

Comparison of caffeine amounts in beverages and other sources

Beverage	Caffeine Content
6-ounce cup of coffee	
drip method	110–150 mg
percolated	64–124 mg
instant	40–108 mg
decaffeinated	2–3 mg
6-ounce cup of hot tea	
1-minute brew	9–33 mg
3-minute brew	20–46 mg
instant	12–28 mg
12-ounce glass of iced tea	22–36 mg
Hot cocoa mix	6 mg
12 ounces of soft drink (cola)	36–46 mg
Some over-the-counter analgesics	32–65 mg
"Stay awake" over-the-counter medications	100–200 mg
Over-the-counter weight-control aids	140–200 mg

SOURCE: Prepared by the staff of Information Plus; data from several sources.

Consumers should be aware that many popular over-the-counter (OTC, or non-prescription) medications contain caffeine. Caffeine is used in some OTC and prescription pain relievers because some studies have shown that caffeine increases their effectiveness. Caffeine is also a major ingredient in most preparations to help individuals stay awake and in OTC weight-control products. (See Table 9.2.)

HOW MUCH CAFFEINE DO WE CONSUME?

The current average daily intake of caffeine in the United States is about 38 milligrams for children ages 5 to 17, the equivalent of one can of soda. Adults consume approximately 200 milligrams on average per day, the equivalent of two cups of coffee. In 1997 Americans consumed 23.5 gallons of coffee per capita.

Results of a study published in 2001 ("Miscalculation of Exposure: Coffee as a Surrogate for Caffeine Intake," *American Journal of Epidemiology*) show that the four main sources of caffeine for persons aged 30 to 75, living in Southern Ontario (Canada), were brewed coffee, instant coffee, regular tea, and cola soft drinks. Brewed coffee was the main source of caffeine. Tea and cola soft drinks ranked second or third. Cola drinks ranked second for those aged 30 to 44, and tea ranked second for those aged 45 to 75. Men had a higher caffeine intake than women, and those aged 45 to 59 had a higher caffeine intake than younger persons.

Results of the 2001 Ontario study also showed that caffeine intake is usually underestimated, because people often report only their coffee intake and omit caffeine from other sources. Researchers found that the average intake of caffeine ranged from 288 mg per day, for women aged 30 to 44, to 426 mg/day for men aged 45 to 59.

According to a 1998 poll by the National Coffee Association of the U.S.A., 7 percent of all 10- to 19-year-olds drink coffee. Some observers predict that coffeehouses will become the teen hangouts of the future. Some observers are concerned about this because the effects of caffeine on teens have not been thoroughly studied.

PHYSICAL EFFECTS OF CAFFEINE

Caffeine is classified as a stimulant because of its effects on the brain, nerves, cardiovascular system, and digestive system. It also has a mild diuretic effect, eliminating water from the body by increasing urination. In moderate amounts, it can help people feel more alert, resulting in their ability to work longer and concentrate better. On the other hand, too much caffeine can produce nervousness, anxiety, irritability, and sleeplessness.

Because it gives athletes a quicker reaction time and more endurance, Olympic competitors are tested for caffeine along with other drugs. If more than a small to moderate amount of caffeine (the amount in a few cups of coffee) is found in the blood, an athlete can be disqualified from the Olympic Games and other international competitions.

The fact that caffeine increases alertness has led to the myth that a strong cup of coffee will help to sober up someone who is drunk, but caffeine does not counteract the effects of alcohol. As an old saying goes, "Giving a drunk person coffee simply means you will have a wide-awake drunk."

Many caffeine-producing plants also contain related stimulants known as methylxanthines, two of which are theophylline and theobromine. Theophylline is used in prescription drugs to treat chronic lung diseases such as asthma and emphysema. Theobromine is a weaker stimulant than caffeine, but it is more abundant in cocoa beans than is caffeine. Part of the stimulating effect of chocolate and cocoa is probably due to theobromine.

How Caffeine Is Metabolized

In its pure state, caffeine is a bitter white powder that looks something like cornstarch. It is a member of the purine family of compounds. When purines are broken down in the body, they form a chemical called xanthine. The liver then converts xanthine to uric acid, a substance found in unusually high levels in human beings. Scientists believe that uric acid contributes to the longer lifespan of humans, as compared to other mammals. Too much uric acid, however, is associated with gout, a painful disease generally affecting the feet and hands.

Caffeine dissolves in water; therefore, after a person drinks or eats something containing caffeine, it spreads to almost all parts of the body. The body takes several hours to metabolize and eliminate caffeine completely from the system. Within three to seven hours, half the caffeine in a drink is eliminated from your body, and the remainder takes the same amount of time to dissipate. Therefore, drinking coffee, tea, or caffeine-containing soft drinks several times a day can build up a large amount of caffeine in the body.

Two other purines, adenine and guanine, play a major role in human genetics and cell function. One of their by-products, adenosine, participates in the supply of energy to cells and helps regulate body processes such as the transmission of signals by nerves. In addition, adenosine can

- Promote sleepiness.

- Dilate blood vessels.

- Reduce the contractions of the stomach and intestines.

- Prevent seizures.

- Slow the reaction to stress.

- Lower the heart rate, blood pressure, and temperature.

To perform these functions, adenosine inhibits the release of neurotransmitters (chemicals that carry messages from one nerve cell to another) by binding to specific receptors on a cell's surface. The structure of caffeine and its by-products is chemically very similar to those of adenosine, which allows caffeine to bind to the same receptor sites, thus blocking the adenosine. As a result, the nerve cells "fire" messages faster and produce the well-known feeling of "caffeine jitters."

IS CAFFEINE ADDICTIVE?

The question of whether caffeine is addictive has been debated for decades. Some medical experts believe that caffeine is mildly addictive because it fits the criteria for addiction: dependence and tolerance. In "Clinical Pharmacology of Caffeine" (*Annual Review of Medicine,* 1990), Dr. Neal L. Benowitz observed, "Minor criteria for addiction liability include the development of tolerance, physical dependence, and recurrent intense desire for the drug, all of which are characteristic of regular caffeine consumers. Thus, there is a group of coffee drinkers who appear to be addicted to caffeine, although the extent of caffeine addiction in the population is unknown."

While caffeine is both psychoactive (mood altering) and addictive, it should be pointed out that, unlike many other drugs, it is not intoxicating. No one gets drunk or "high" on caffeine, although it can produce anxiety and sleeplessness.

A research team from Johns Hopkins University School of Medicine in Baltimore, Maryland, examined the daily intake of caffeine of 16 people who thought that they had problems with caffeine. The results of the study, reported in the *Journal of the American Medical Associa-*

tion (Eric C. Strain, et al., "Caffeine Dependence Syndrome: Evidence from Case Histories and Experimental Evaluations," October 5, 1994), indicated that these people showed signs of dependence on caffeine.

Some of the persons in the study had been advised by their doctors to give up caffeine but had been unable to do so. One person averaged only 129 milligrams of caffeine a day (equivalent to one cup of moderately strong coffee), while others drank more than 1,000 milligrams per day. Only eight of the people studied got their caffeine from coffee. One person drank tea, and the other seven drank soft drinks.

When caffeine was removed for two days from the diets of 11 people in the study, nine of them reported withdrawal symptoms. Seven persons had severe headaches. Other symptoms included fatigue, depression, and impaired performance of daily work or chores. The researchers concluded that those who take caffeine can become physically dependent on the drug.

However, results of a 1999 study, "Are We Dependent Upon Coffee and Caffeine? A Review on Human and Animal Data" (*Neuroscience Biobehavior Review*), show that complete tolerance to many of the effects of caffeine on the central nervous system does not occur. Additionally, average daily doses of caffeine do not act on brain structures related to reward, motivation, and addiction as do "hard" drugs such as cocaine and amphetamines.

HEALTH EFFECTS

One major side effect of consuming coffee, tea, soft drinks, and chocolate is the nervous feeling or jitters they can cause when an individual consumes too much. In general, the more caffeine that is consumed, the more likely the person is to suffer these side effects. For some people, however, even a minor amount of caffeine can disrupt their sleeping patterns, cause their hands to shake, or create anxiety.

The physical effects of caffeine in the body are fairly well understood, but studies attempting to link caffeine to different diseases have been inconclusive. For example, scientists know that caffeine stimulates the central nervous system, resulting in increased alertness and mood elevation. However, for persons in good general health, studies have not shown any link between drinking coffee or tea and suffering from a nervous disorder, other than the temporary side effects discussed above.

High doses of caffeine (from 200 to 300 mg) may cause the heart to beat harder and faster and can raise blood pressure slightly, especially in persons not used to consuming caffeine. For those who consume caffeine on a regular basis, blood pressure returns to a baseline level. A review of medical literature indicates that moderate caf-

TABLE 9.3

Effects of caffeine on certain health problems

Health Problem	Caffeine Effects
High blood pressure	Caffeine raises blood pressure.
Ulcers	Caffeine increases stomach acid. Coffee irritates the stomach lining.
Diabetes or hypoglycemia (low blood sugar)	Caffeine speeds up metabolism slightly, which can change blood sugar levels unpredictably.

SOURCE: "Table 9.3: Effects of Caffeine on Certain Health Problems," Prepared by staff of Information Plus.

feine consumption does not cause high blood pressure or any increased incidence of coronary heart disease.

Coffee has also been suspected of increasing the cholesterol level in the blood. Recent medical studies using filtered coffee, however, have found little or no evidence of a relationship between caffeine and high cholesterol levels. Brewing methods may make a difference—studies from Scandinavia using boiled unfiltered coffee did find a relationship. Nevertheless, consumption of coffee as typically prepared in the United States does not seem to affect blood cholesterol levels.

Caffeine Overdose

It is possible to overdose on caffeine. Symptoms of an overdose include hyperventilation (rapid, deep breathing), very fast heartbeat, heart fibrillation (twitching; beating unevenly), convulsions, and imbalances in the body's levels of potassium, sugar, and other blood chemicals. The overdose amount differs from person to person, depending on body weight, metabolism, and usual intake of caffeine (tolerance).

Fatal caffeine overdoses are very rare, but they can occur. Most are either suicide attempts or accidental overuse of caffeine tablets. An average-sized adult would have to consume about 5,000 milligrams (5 grams) of caffeine—about 50 cups of normal-strength coffee—to get a fatal dose, and the beverages would have to be drunk in rapid succession. In children, because of their smaller body mass and differences in metabolizing caffeine, a much smaller dose could be lethal.

Is There a Link to Cancer?

Studies attempting to show a direct link between caffeine and various types of cancer have been contradictory. For every study that gives evidence of a possible link, another study finds no evidence of a link. Results of scientific experiments have shown that caffeine can change the cells of bacteria, plants, insects, and humans in the laboratory, but results of studies have not shown that caffeine causes cancerous cell development in generally healthy persons. In 1997 the American Institute of Cancer

Research (Washington, D.C.) reported that "most evidence suggests that regular consumption of coffee and/or tea has no significant relationship with the risk of cancer of any site."

Caffeine and Women's Health

Studies of the effects of caffeine on women's health have also been conflicting or inconclusive. Some early studies indicated that caffeine contributed to fibrocystic disease (noncancerous growths in the breasts or uterus), but several more recent studies, including a 1986 study of more than 3,400 women conducted at the National Cancer Institute, found no direct association.

One study indicated that drinking more than three cups of coffee a day lowers a woman's chance of conception by 27 percent. A study of large numbers of pregnant women found that women who drank coffee regularly were more likely to give birth to low birthweight babies than were women who did not drink coffee. The medical community is still debating these findings.

While heavy caffeine use has been suspected of contributing to miscarriages, recent studies gave conflicting results and failed to establish a definite link. However, results of a study published in 2000 in the *New England Journal of Medicine* ("Caffeine Intake and the Risk of First-Trimester Spontaneous Abortion") suggest that the ingestion of caffeine may increase the risk of an early spontaneous abortion among nonsmoking women carrying genetically normal fetuses.

The U.S. Food and Drug Administration (FDA) advises pregnant women to limit or eliminate their intake of caffeine. The substances that an expectant mother eats or drinks can be passed through the bloodstream to her unborn baby, and a mother who drinks large amounts of caffeine may produce a baby who shows signs of caffeine jitters. Nursing mothers can also pass caffeine to their babies through breast milk, leading to sleep disruptions and irritability.

Effects on People with Known Health Problems

While caffeine may not be a direct cause of disease, it can aggravate some already existing health problems. Persons with high blood pressure, diabetes, hypoglycemia (low blood sugar), or ulcers should consult their doctors regarding caffeine use. Table 9.3 shows the interaction of caffeine with these conditions.

Caffeine may also interact with prescription and OTC drugs. In some cases, a drug may have side effects similar to those of caffeine, so a lower amount of coffee, tea, or cola may cause jitteriness or disturb sleep. In other cases, caffeine may reduce or heighten the effectiveness of medications. In addition, alcohol and tobacco affect the metabolism of caffeine in the body, causing it either to leave the body more rapidly or to remain in the bloodstream for a longer time. (See Table 9.4.) People who drink a lot of caffeinated beverages should discuss with their doctors the effects caffeine might have on any medications they are taking.

Possible Health Benefits

Caffeine's bronchodilator effect of opening the airways can be beneficial to persons suffering from acute bronchial asthma. Caffeine acts as an analgesic (pain killer) and, when combined with ibuprofen, can bring faster and longer lasting relief from tension headaches. It is often prescribed for migraine headaches. Drinking two cups of coffee before breakfast has helped individuals suffering from dizziness and other effects caused by abnormally low blood pressure.

Coffee has been associated with a reduced risk of colon cancer, and tea has been related to a reduced risk of rectal cancer. New data strongly suggest that caffeine has an antidepressant effect. Michael F. Leitzmann, et al., in "A Prospective Study of Coffee Consumption and the Risk of Symptomatic Gallstone Disease in Men" (*Journal of the American Medical Association,* June 9, 1999), reported that men who drank two to three cups of coffee a day had a 40 percent lower risk of gallstones than those who did not drink regular coffee.

DECAFFEINATION

People who are sensitive to caffeine or those who want to moderate their caffeine intake can still enjoy their favorite beverages in decaffeinated forms. Decaffeination removes almost all the caffeine (about 97 percent) from coffee or tea. There are three common methods of decaffeination.

- The direct method uses methylene chloride, which dissolves caffeine, to remove caffeine from green coffee beans and moist tea leaves. Then a further process removes the methylene chloride from the product.

- The water method uses water to dissolve the caffeine from steamed coffee beans or tea leaves. Then the caffeine is removed from the water by using methylene chloride, and the decaffeinated water is added back to the coffee or tea to enhance flavor. The beans or leaves are steamed again to remove any methylene chloride remaining.

TABLE 9.4

Caffeine interactions with other substances

Substance	Interaction
Alcohol	Slows down the metabolism of caffeine
Tobacco	Speeds up the metabolism of caffeine
Birth control pills	Slow down the metabolism of caffeine
Monoamine oxidase (MAO) inhibitors (used to treat depression) and some tranquilizers	May have reduced effectiveness due to caffeine's stimulating effects
Decongestants	Have effects similar to caffeine, so jitteriness, insomnia, or irritability may be made worse

SOURCE: Prepared by the staff of Information Plus; data from several sources.

- The carbon dioxide method, which is the newest, uses carbon dioxide to remove caffeine. This process leaves in more flavor than the other two methods.

Methylene chloride, in its inhaled form, is known to cause cancer in laboratory animals. When methylene chloride is diluted in their drinking water, however, the laboratory animals do not develop cancer. The FDA has analyzed decaffeinated coffee made through the direct method and found that so little methylene chloride is left in the final product that any risk of it causing cancer is remote. The amount of methylene chloride in the finished product is less than that found in the air on a smoggy day.

Some people find that decaffeinated coffee lacks some of the flavor of regular coffee. Using fresh-ground or flavored decaffeinated coffee beans may help brew a better tasting pot of coffee. Using half regular coffee and half decaffeinated coffee can also help cut down caffeine intake, as can simply drinking fewer caffeine-laden drinks per day. Hot chocolate, hot apple cider, and herbal teas make good low-caffeine or caffeine-free substitutes for those wanting to limit their caffeine intake.

What Do They Do with the Caffeine They Take Out?

The caffeine removed from coffee and tea is a useful by-product. It can be used in medications or added to soft drinks to increase their caffeine content. Some is processed into theophylline, which is used in medications for asthma and emphysema.

CHAPTER 10
ATTITUDES TOWARD ALCOHOL AND TOBACCO

PUBLIC OPINIONS ON DRINKING ALCOHOL

Most Americans Drink Alcohol

A 2000 Gallup Poll found that 64 percent of Americans drink alcoholic beverages, such as liquor (distilled spirits), wine, or beer, up from 58 percent in 1996. Thirty-six percent were "teetotalers," or total abstainers. Over the past 15 years or so, the levels of those who drank alcoholic beverages remained below the 71 percent the Gallup Poll recorded in 1976, 1977, and 1978. (See Table 10.1.)

Beer, preferred by 43 percent of America's drinkers, was the most popular alcoholic beverage in 2000, followed by wine and then liquor. In 1999 the popularity of wine (preferred by 34 percent of respondents that year) was at its highest level in recent years, but dropped to 31 percent in 2000. Preference for liquor (22 percent) was up from the 1999 level of 19 percent. (See Table 10.2.)

One-fourth (26 percent) of drinkers had consumed an alcoholic beverage within the past 24 hours, and one-third (32 percent) had ingested alcohol within the past week. A significantly smaller proportion of people had taken a drink in the past 24 hours in 2000 than in 1984, and a slightly larger proportion had taken a drink between one day and a week ago. Especially notable is the fact that 42 percent of drinkers in 2000 reported that it had been more than a week since they had taken a drink, compared to 31 percent of the 1984 drinkers. (See Table 10.3.)

In 2000 most respondents claimed that they had either abstained from drinking or had been light-to-moderate drinkers during the past week. More specifically, 43 percent of the drinkers polled reported that they had not drunk an alcoholic beverage within the past week, and 46 percent had drunk seven or fewer drinks. (See Table 10.4.) Most drinkers (74 percent) did not think they drank more than they should, although a significant proportion (26 percent) reported that they sometimes drank more than they felt they should (see Table 10.5).

TABLE 10.1

Percentage of responses to the question "Do you have occasion to use alcoholic beverages such as liquor, wine, or beer, or are you a total abstainer?"

	Yes, drink %
2000 Nov 13–15*	64
1999 Sep 23–26	64
1997 Jun 26–29	61
1996 Jun 27–30	58
1994 Jun 3–6	65
1992	64
1990	57
1989	56
1988	63
1987	65
1984	64
1983	65
1982	65
1981	70
1979	69
1978	71
1977	71
1976	71
1974	68
1969	64
1966	65
1964	63
1960	62
1958	55
1957	58
1956	60
1952	60
1951	59
1950	60
1949	58
1947	63
1946	67
1945	67
1939	58

*2000 data based on telephone interviews with a randomly selected national sample of 1028 persons, 18 years and older. (Error margin plus or minus 4 percentage points.)

SOURCE: Wendy W. Simmons, "Do you have occasion to use alcoholic beverages such as liquor, wine or beer, or are you a total abstainer?" in *One in Six Americans Admit Drinking Too Much,* Gallup News Service, Princeton, NJ, December 4, 2000

TABLE 10.2

Percentage of responses to the question "Do you most often drink liquor, wine, or beer?"

	Liquor %	Wine %	Beer %	ALL/SAME (vol.) %	OTHER (vol.) %	No opinion %
2000 Nov 13–15	22	31	43	3	0	1
1999 Sep 23–26	19	34	42	4	*	1
1997 Jun 26–29	18	32	45	4	*	1
1996 Jul 25–28	20	27	46	6	0	1
1994 Jun 3–6	18	29	47	3	1	2
1992 Jan	21	27	47	3	1	1

*means less than 0.5%.
(vol.) means volunteered response.
2000 data based on 692 respondents who drink. (Error margin: plus or minus 4 percentage points)

SOURCE: Wendy W. Simmons, "Do you most often drink liquor, wine, or beer?" in *One in Six Americans Admit Drinking Too Much,* Gallup News Service, Princeton, NJ, December 4, 2000

TABLE 10.3

Percentage of responses to the question "When did you last take a drink of any kind of alcoholic beverage?"

	24 hours %	Over 1 day to week ago %	Over 1 week ago %	Don't know %
2000	26	32	42	*
1999	35	25	39	1
1997	33	25	42	*
1996	28	26	45	1
1994	34	23	42	1
1992	26	24	49	1
1990	29	23	47	1
1989	32	35	32	1
1988	39	25	34	2
1987	38	30	31	1
1984	39	29	31	1

*means less than 0.5%.

SOURCE: Wendy W. Simmons, "When did you last take a drink of any kind of alcoholic beverage?" in *One in Six Americans Admit Drinking Too Much,* Gallup News Service, Princeton, NJ, December 4, 2000

Family Problems Caused by Alcohol

According to the 2000 Gallup survey, drinking caused family problems for nearly four out of ten Americans. This is the highest rate recorded since Gallup first asked the question in 1947. The proportion of respondents in 2000 who thought drinking had sometimes created a problem in their families (36 percent) was considerably higher than the proportion of drinkers (26 percent) who thought that they sometimes drank too much. (See Table 10.5 and Table 10.6.) This discrepancy could be due to changes in drinking patterns or could reflect a decline in the social acceptability of admitting to drinking problems at the same time that willingness to report family problems has increased.

Concern About Underage Drinking

According to the 1998 *Youth Access to Alcohol Survey,* commissioned by the Robert Wood Johnson Foundation (Princeton, New Jersey), the nation's largest philanthropic organization supporting health and health care, Americans view underage drinking as a significant problem and support a variety of measures to help reduce teen drinking. Ninety-six percent said they were very or somewhat concerned about teen drinking.

Nancy Kaufman, vice president of the Robert Wood Johnson Foundation, maintains that

> Underage drinking is a vast problem with grave consequences. It is a factor in nearly half of all teen automobile crashes, the leading cause of death among teens. Beyond that, alcohol contributes to suicides, homicides, and fatal injuries and is a factor in sexual assaults and date rapes. Obviously, something needs to be done to avoid these serious problems.

Table 10.7 ranks 26 suggested policies to reduce teenage drinking, from most to least supported among all respondents. The table also lists the subgroups found to be the most and least supportive of each policy item. The highest percentages of support were for policies that deal with restrictions on drinking alcoholic beverages in various public locations. For instance, 92 percent of respondents would favor restrictions on drinking on city streets, and 89 percent supported restrictions on drinking on college campuses.

At the bottom of the list, with 40 percent or less support, are policies banning certain types of alcohol sales. Forty percent of the survey respondents would ban happy hours; only 31 percent would support a ban on beer keg sales to individuals. (See Table 10.7.)

Support for alcohol tax increases depended on the use for the revenues gained. If the taxes were used for prevention purposes, 82 percent would favor such a tax. Fewer (70 percent) supported a tax increase if it were used to lower other taxes, such as income taxes. Only 37 percent favored a tax increase if the revenues were used for "any government purpose." (See Table 10.7.)

Several questions concerned alcohol advertising, on which more than half of the respondents favored restrictions. Sixty-seven percent opposed the use of cartoons or youth-oriented materials on alcoholic beverage packaging. Two-thirds (67 percent) supported a ban on liquor ads on TV, while 61 percent favored a ban on beer or wine TV ads. Sixty-three percent would outlaw alcohol billboard ads, and nearly 6 out of 10 (59 percent) favored banning the use of sports teams and athletes as symbols in alcohol marketing. (See Table 10.7.)

Women appeared to be most supportive of restrictive alcohol policies. Male respondents and those aged 18 to 24 were more likely to indicate lower levels of support for restrictive alcohol policies. The policy that revealed the largest difference of opinion was the restriction of

TABLE 10.4

TABLE 10.5

Percentage of responses to the question "Approximately how many drinks of any kind of alcoholic beverages did you drink in the past seven days?"

	0 %	1-7 %	8-19 %	20+ %	Don't know %
2000 Nov 13–15	43	46	8	3	*
1999	40	47	8	4	*
1997	41	45	8	5	1
1996	47	42	7	2	2
1994	44	42	10	3	1
1992	51	36	10	2	1
1990	50	40	6	3	1
1989	33	47	13	5	2
1988	32	49	10	6	3
1987	29	50	11	6	4

*means less than 0.5%.
2000 data based on 692 respondents who drink. (Error margin: plus or minus 4 percentage points)

SOURCE: Wendy W. Simmons, "Approximately how many drinks of any kind of alcoholic beverages did you drink in the past seven days?" in *One in Six Americans Admit Drinking Too Much,* Gallup News Service, Princeton, NJ, December 4, 2000

Percentage of responses to the question "Do you sometimes drink more alcoholic beverages than you think you should?"

	Yes %	No %	No opinion %
2000	26	74	*
1999	24	76	*
1997	22	78	*
1996	25	75	*
1994	29	71	0
1992	29	71	—
1990	23	76	1
1989	35	65	—
1987	29	71	—
1985	32	68	—
1978	23	77	—

*means less than 0.5%.
2000 data based on 692 respondents who drink. (Error margin: plus or minus 4 percentage points)

SOURCE: Wendy W. Simmons, "Do you sometimes drink more alcoholic beverages than you think you should?" in *One in Six Americans Admit Drinking Too Much,* Gallup News Service, Princeton, NJ, December 4, 2000

TABLE 10.6

Percentage of responses to the question "Has drinking ever been a cause of trouble in your family?"

	Yes %	No %	No answer %
2000	36	64	*
1999	36	64	*
1997	30	70	*
1996	23	77	*
1994	27	72	1
1992	24	76	—
1990	23	76	1
1989	19	81	—
1987	24	76	—
1985	21	79	—
1984	18	82	—
1981	22	78	—
1978	22	78	—
1976	17	83	—
1974	12	88	—
1966	12	88	—
1947	15	85	—

*means less than 0.5%.
2000 data based on 692 respondents who drink. (Error margin: plus or minus 4 percentage points)

SOURCE: Wendy W.Simmons, "Has drinking ever been a cause of trouble in your family?" in *One in Six Americans Admit Drinking Too Much,* Gallup News Service, Princeton, NJ, December 4, 2000

drinking at sports stadiums. There was a difference of 23 points between the responses of women and those aged 18 to 24. (See Table 10.7.)

Nearly three-quarters (73 percent) of respondents supported a "zero-tolerance" policy for young drivers, in which teenagers would be punished if they tested positive for any amount of alcohol in their blood. Women (80 percent) were the most supportive, with men (66 percent) less supportive. (See Table 10.7.)

Seven out of ten respondents believed that stiffer punishments for teenagers caught drinking would discourage them from obtaining alcohol (see Figure 10.1). Most (51.7 percent) favored a one-year license suspension, while 34.5 percent supported a penalty of 20 hours of community service. The two other survey choices had less support—only 10.5 percent advocated a $500 fine, and just 3.3 percent backed ineligibility for future state college scholarships and loans. (See Figure 10.2.)

The *Youth Access to Alcohol Survey* also indicated that respondents believe adults providing alcohol to teens, as well as the teens themselves, are responsible for problems associated with teen drinking. Eighty-three percent supported penalties for adults who provide alcohol to underage drinkers. (See Figure 10.3.)

PUBLIC OPINIONS ON SMOKING

How Many Smoke?

While most Americans do not smoke, about one-quarter of them do. A 2000 Gallup survey on smoking reported that 25 percent of the adults interviewed said they had smoked cigarettes during the past week, down from 40 percent in 1969 and 45 percent in 1954. (See Table 5.2 in Chapter 5.) Most smokers consumed one pack (29 percent) or less (62 percent) a day, while 9 percent smoked more than one pack per day, down from 27 percent in 1977. (See Table 5.3 in Chapter 5.)

The percentage of Americans of various age groups who smoke has been quite stable over the past decade. In

TABLE 10.7

Rank order of public support for alcohol policies wth most and least supportive groups

% of Overall Support	Policy Type	Supportive group: Most	Least	Range[1]
92%	Restrict drinking on city streets	Women	18-24	9
90%	Restrict drinking at parks	Women	Liberals	9
89%	Restrict drinking on college campuses	Women	18-24	12
89%	Require server training	Liberals	Conservatives	4
88%	Require bar owner training	Liberals	Conservatives	6
85%	Restrict drinking at concerts	Women	18-24	15
83%	Punish adult providers	Republicans	18-24	16
82%	Restrict drinking at beaches	Women	18-24	15
82%	Tax increase for prevention purposes	Women & Democrats	Men	7
80%	Require legal age for alcohol servers	25+, Democrats & Conservatives	18-24	10
77%	Restrict drinking at sports stadiums	Women	18-24	23
73%	Zero tolerance for youth (BAC 0.00)	Women	Men	14
70%	Tax increase for tax relief	18-24	Men	18
67%	Ban youth-oriented packaging	Women	Men	14
67%	Ban liquor ads on TV	Women	Men	12
66%	Compliance checks at liquor stores	Republicans	Democrats	7
66%	Allow local controls on alcohol	Conservatives	Liberals & 18-24	11
63%	Ban alcohol billboard ads	Women	Men & 18-24	14
61%	Ban beer/wine ads on TV	Women	Men & 18-24	17
61%	Require beer keg registration	Women	Men	12
59%	Ban home delivery of alcohol	Republicans	Men	8
59%	Ban alcohol marketing with athletes	Women	Men	16
40%	Ban happy hours	Republicans	Liberals	13
40%	State control of liquor sales	18-24	Republicans	16
37%	Tax increase for any government purpose	18-24	Men	18
31%	Ban beer keg sales to individuals	Women & Conservatives	Men & Liberals	12

[1]Range is calculated by subtracting the lowest from the highest percent to indicate the spread or distance between most and least supportive groups.

SOURCE: E.M. Harwood, A.C. Wagenaar, and K.M. Zander "Table 10.7: Rank order of public support for alcohol policies with most and least supportive groups," in *Youth Access to Alcohol Survey: Summary Report*, University of Minnesota, Minneapolis, MN, 1998

FIGURE 10.1

Percentage of various responses to the statement "Stiffer punishments for teenagers who are caught drinking will discourage them from getting alcohol."

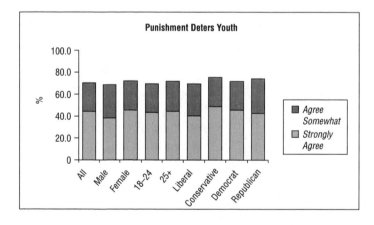

% who Agree that Punishment Deters Youth

	All	Male	Female	18–24	25+	Liberal	Conservative	Democrat	Republican
Strongly Agree	44.4	40.7	47.7	44.1	44.7	38.9	50.4	46.8	43.1
Agree Somewhat	26.6	28.1	25.4	26.9	26.7	28.6	24.7	23.6	30.3
Disagree Somewhat	14.3	15.2	13.5	15.4	13.9	13.7	12.5	13.1	16.0
Strongly Disagree	14.7	16.1	13.5	13.6	14.7	18.8	12.4	16.5	10.7

SOURCE: E.M. Harwood, A.C. Wagenaar, and K.M. Zander, *Youth Access to Alcohol Survey: Summary Report*, University of Minnesota, Minneapolis, MN, 1998

FIGURE 10.2

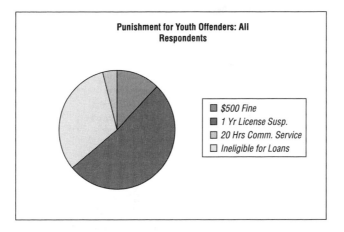

Percentage of various responses to the question "If a teenager is caught drinking, which of the following do you feel is the most appropriate punishment?"

Possible responses: A fine of $500, Drivers license suspended for one year, Twenty hours of community service, Not eligible for future state college scholarships and loans.

Punishment for Youth Offenders: All Respondents

- $500 Fine
- 1 Yr License Susp.
- 20 Hrs Comm. Service
- Ineligible for Loans

% in favor of each punishment for youth offenders

	All	Male	Female	18–24	25+	Liberal	Conservative	Democrat	Republican
$500 Fine	10.5	12.2	9.0	16.1	9.6	13.0	9.2	8.9	12.7
1 yr license susp.	51.7	49.0	54.0	41.6	53.6	46.2	55.5	53.0	51.9
20 hrs comm. service	34.5	36.1	33.1	37.6	33.8	39.5	32.4	35.1	31.7
Ineligible for loans	3.3	2.7	3.8	4.7	3.1	1.3	2.9	3.1	3.8

SOURCE: E.M. Harwood, A.C. Wagenaar, and K.M. Zander, *Youth Access to Alcohol Survey: Summary Report*, University of Minnesota, Minneapolis, MN, 1998

2000, as Table 10.8 shows, between 26 and 30 percent of adults ages 18 to 64 had smoked in the past week. The proportion of smokers drops dramatically to about 15 percent in those over age 65. And as Table 10.9 reveals, most smokers who responded to the 2000 survey (76 percent) began their habit prior to or at age 18.

How Many Want to Quit?

Eight out of ten (82 percent) of the smokers polled in 2000 reported that they would like to give up smoking, up from 76 percent in 1999. (See Table 10.10.) Many (65 percent) claimed they had made at least one serious effort to stop smoking, a significant decrease from 74 percent in 1996. (See Table 10.11.) It appears that many smokers wish to quit smoking but take no action to do so, although in 1999, 85 percent of smokers said they would not start smoking if they had it to do all over again. (See Table 10.12.) Seventy-four percent of year 2000 smokers believed they were addicted to cigarettes, up significantly from 61 percent in 1990. (see Table 10.13)

Should Smoking Be Limited?

The proportion of those who consider secondhand smoke harmful has increased since 1994. In that year, the Gallup Poll reported that 78 percent of respondents felt that

secondhand smoke was somewhat harmful (42 percent) or very harmful (36 percent). In 1999 the proportion who felt that it was somewhat (39 percent) or very (43 percent) harmful had grown to 82 percent, an increase of 4 percent over five years. (See Table 5.9 in Chapter 5.) Nevertheless, 55 percent of the 1997 respondents felt that secondhand smoke is very harmful, while only 43 percent of the 1999 respondents held this opinion.

In its 2000 survey of smoking, the Gallup Poll found that most Americans supported some restrictions on smoking. Most of these restrictions were expressed in support of limiting smoking to designated areas. For example, 65 percent believed smoking should be limited to designated areas in hotels and motels, although a significant percentage thought smoking should be banned in these areas (28 percent). (See Table 10.14.)

As regards the workplace, 57 percent of those polled believed smoking should be permitted only in designated areas in the workplace, while 37 percent thought smoking should be banned. About half (48 percent) wanted smoking limited to designated areas in restaurants, but most of the other half (47 percent) wanted it banned. Very few thought there should be no restrictions at all on smoking. Looking at Gallup's figures between 1994 and 2000, the proportion of respondents wanting no restrictions has

FIGURE 10.3

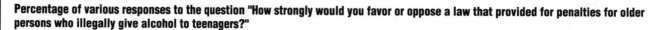

Percentage of various responses to the question "How strongly would you favor or oppose a law that provided for penalties for older persons who illegally give alcohol to teenagers?"

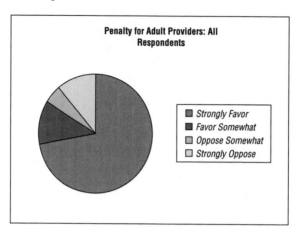

Penalty for Adult Providers: All
Respondents

- Strongly Favor
- Favor Somewhat
- Oppose Somewhat
- Strongly Oppose

% in Favor of Penalty for Adult Providers

	All	Male	Female	18–24	25+	Liberal	Conservative	Democrat	Republican
Strongly Favor	70.9	67.2	74.1	49.0	74.2	64.8	73.0	71.2	73.0
Favor Somewhat	12.3	15.2	9.8	21.9	11.0	15.1	10.5	11.0	13.5
Oppose Somewhat	5.4	6.9	4.2	13.7	4.3	7.8	4.2	3.9	5.0
Strongly Oppose	11.4	10.8	11.9	15.5	10.4	12.3	12.4	13.9	8.5

SOURCE: E.M. Harwood, A.C. Wagenaar, and K.M. Zander, *Youth Access to Alcohol Survey: Summary Report*, University of Minnesota, Minneapolis, MN, 1998

remained somewhat stable, while the percentage wanting a total ban has increased. (See Table 10.14.)

Responsibility for Health Damages

In 1998 the major tobacco companies settled with state attorneys general to reimburse states for the Medicaid costs of treating smokers. Public health organizations and others have long claimed that the cigarette industry should be held legally and financially responsible for the health problems of those who use its products. In the past, the tobacco industry had defended itself by denying that health hazards had been proved by scientific testing, and by asserting that using tobacco is a matter of choice.

In 1991, when the Gallup Organization first polled Americans on the topic of liability, only 13 percent of respondents believed that tobacco companies should be held legally responsible for health damages related to smoking. Sixty-six percent felt that the government-required warnings on tobacco packaging absolved the companies from responsibility. The 1996 poll, however, reported a much different picture. Among those responding, 30 percent felt the companies should be held accountable for health damages, and only half (51 percent) still considered the package warnings sufficient to excuse the companies. In polls taken in 1997, 1999, and 2000, the

picture remained substantially the same as in 1996: 25 to 30 percent of respondents thought that the cigarette companies were completely or mostly to blame. Over half, 55 to 64 percent, thought that smokers were mostly or completely to blame. (See Table 10.15.)

In the week following the 1998 tobacco settlement, the Gallup Organization asked Americans their opinions of the settlement. At the time, only about half of the respondents said they had heard much about it. When asked if they favored or opposed the agreement, over one-third (35 percent) of the respondents either said they had not heard about it or had no opinion on the tobacco agreement. More of the respondents who had heard about the agreement and had an opinion favored (40 percent) rather than opposed (25 percent) the agreement. One-third of the smokers felt the settlement was too tough on the tobacco companies, while one-quarter felt it was not tough enough. Thirty-nine percent of nonsmokers thought it was not tough enough.

Should Tobacco Be a Controlled Substance?

An important issue in the debate over cigarettes has been the question of government regulation. The federal government contends that tobacco is a drug and should be regulated by the Food and Drug Administration (FDA). The tobacco industry, not surprisingly, strongly opposes such a move. On March 21, 2000, the Supreme Court

TABLE 10.8

Percentage of Americans of various age groups who smoked cigarettes in the past week

	18-29	30-49	50-64	65+
2000	30%	26	29	15
1989	28%	30	31	16

Note: 2000 data based on telephone interviews with a randomly selected national sample of 1,028 adults, 18 years and older. Error margin is plus or minus 3 percentage points.

SOURCE: Lydia Saad, "Percent Who Smoked Cigarettes in Past Week" in *Smoking in Restaurants Frowned on by Many Americans,* Gallup News Service, Princeton, NJ, November 29, 2000

TABLE 10.10

Percentage of responses to the question "All things considered, would you like to give up smoking, or not?"

	Yes %	No %	No opinion %
2000	82	16	2
1999	76	23	1
1996	73	26	1
1994	70	28	2
1991	76	22	2
1990	74	24	2
1988	68	27	5
1986	75	22	3
1981	66	30	4
1977	66	29	5

2000 data based on 239 respondents who smoke. (Error margin: plus or minus 7 percentage points)

SOURCE: Lydia Saad, "All things considered, would you like to give up smoking or not?" in *Smoking in Restaurants Frowned on by many Americans,* Gallup News Service, Princeton, NJ, November 29, 2000

TABLE 10.12

Question: If you had to do it all over again, would you start smoking, or not?

[Based on—226—smokers; Margin of error ± 7 PCT points]

	Jul 6-8 1990	Sep 23-26 1999
Yes	13%	13%
No	83	85
No opinion	4	2
	100%	100%

SOURCE: David W. Moore, "If you had to do it all over again, would you start smoking, or not?" in *Nine of Ten Americans View Smoking as Harmful,* Gallup News Service, Princeton, NJ, October 7, 1999

TABLE 10.9

Percentage of various responses to the question "At what age did you begin smoking?"

	Under 16 %	16-18 %	Over 18 %	No opinion %	Mean age
2000	37	39	21	3	17
1999	36	35	29	*	18
1994	32	37	29	2	—
1991	34	36	29	1	—

*Less than 0.5%
Note: 2000 data based on telephone interviews with 239 respondents who smoke. Error margin is plus or minus 7 percentage points.

SOURCE: Lydia Saad, "At what age did you begin smoking?" in *Smoking in Restaurants Frowned on by Many Americans,* Gallup News Service, Princeton, NJ, November 29, 2000

TABLE 10.11

Question: Have you ever made a really serious effort to stop smoking or not?

[Based on 226 smokers; Margin of error ± 7 percentage points]

	Yes	No/Don't Know
1999 Sep 23–26	65%	35%
1997 Jun 26–29	62	38
1997 Jun 23–24	59	41
1996	74	26
1994	68	32
1991	64	35
1990	67	33
1989	60	40
1988	70	30

SOURCE: David W. Moore, "Have your ever made a really serious effort to stop smoking or not?" in *Nine of Ten Americans View Smoking as Harmful,* Gallup News Service, Princeton, NJ, October 7, 1999

TABLE 10.13

Percentage of responses to the question "Do you consider yourself addicted to cigarettes or not?"

	Yes, addicted %	No, not %	No opinion %
2000	74	26	0
1999	72	28	*
1996	69	31	0
1991	70	29	1
1990	61	39	*

*means less than 0.5%

2000 data based on 239 respondents who smoke. (Error margin: plus or minus 7 percentage points)

SOURCE: Lydia Saad, "Do you consider yourself addicted to cigarettes or not?" in *Smoking in Restaurants Frowned on by many Americans,* Gallup News Service, Princeton, NJ, November 29, 2000

ruled (5-4) that the government lacks authority to regulate tobacco as an addictive drug.

In 1998 a Fox News/Opinion Dynamics Poll surveyed Americans about government regulations on the availability of cigarettes. Half of all respondents (51 percent) felt the government should not regulate availability. Not surprisingly, 70 percent of smokers were against regulation.

TABLE 10.14

Percentage of responses to the question "What is your opinion regarding smoking in public places?"

	1987	1990	1991	1994	1999	2000
Hotels and Motels						
Set aside areas	67%	73%	70%	68%	70%	65%
Totally ban	10	18	17	20	24	28
No restrictions	20	8	12	10	6	7
Workplaces						
Set aside areas	70	69	67	63	61	57
Totally ban	17	25	24	32	34	37
No restrictions	11	5	8	4	4	6
Restaurants						
Set aside areas	74	66	66	57	56	48
Totally ban	17	30	28	38	40	47
No restrictions	8	4	5	4	4	5

*means less than 0.5%

Note: "No opinion" omitted.

2000 data based on telephone interviews with a randomly selected national sample of 1,028 persons, 18 years and older. (Error margin: plus or minus 3 percentage points)

SOURCE: Lydia Saad, "What is your opinion regarding smoking in public places? First in [ROTATED]—should they SET ASIDE certain areas, should they totally BAN smoking, or should there be NO RESTRICTIONS on smoking? How about in…" in *Smoking in Restaurants Frowned on by many Americans,* Gallup News Service, Princeton, NJ, November 29, 2000

When asked if they thought cigarettes should be illegal, almost one-fourth (23 percent) of all respondents favored making cigarettes illegal. Eighty-five percent of smokers were against making cigarettes illegal.

TABLE 10.15

Percentage of responses to the question "Which of the following statements best describes your view of who's to blame for the health problems faced by smokers in this country?"

	Companies completely to blame %	Companies mostly to blame %	Equally to blame (vol.) %	Smokers mostly to blame %	Smokers completely to blame %	No opinion %
2000	6	23	8	35	27	1
1999	9	21	13	31	24	2
1997	5	20	10	38	26	1

2000 data based on telephone interviews with a randomly selected national sample of 1,028 persons, 18 years and older. (Error margin: plus or minus 3 percentage points) (vol.) means volunteered response

SOURCE: Lydia Saad, "Which of the following statements best describes your view of who's to blame for the health problems faced by smokers in this country? {[ROTATE 1-4/4-1] 1) The tobacco companies are completely to blame, 2) The tobacco companies are mostly to blame, 3) Smokers are mostly to blame, 4) Smokers are completely to blame?}" in *Smoking in Restaurants Frowned on by many Americans,* Gallup News Service, Princeton, NJ, November 29, 2000

In 1999 the Gallup Organization asked smokers what they would be most likely to do if the government made cigarettes illegal. Nearly two-thirds (63 percent) said they would quit, while one-third (32 percent) said they would continue to smoke by trying to get cigarettes illegally.

IMPORTANT NAMES AND ADDRESSES

AAA Foundation for Traffic Safety
1440 New York Ave. NW, #201
Washington, DC 20005
(202) 638-5944
FAX: (202) 638-5943
http://www.aaafts.org

Action on Smoking and Health (ASH)
2013 H St. NW
Washington, DC 20006
(202) 659-4310
http://www.ash.org

Al-Anon Family Group Headquarters
1600 Corporate Landing Pkwy.
Virginia Beach, VA 23454-5617
(888) 4AL-ANON
http://www.al-anon.alateen.org

Alcoholics Anonymous World Services
475 Riverside Dr., 11th Floor.
New York, NY 10115
Grand Central Station
P.O. Box 459
New York, NY 10163
(212) 870-3400
FAX: (212) 870-3003
http://www.aa.org

American Association for World Health
1825 K St. NW, #1208
Washington, DC 20006
(202) 466-5883
FAX: (202) 466-5896
http://www.aawhworldhealth.org

Beer Institute
122 C St. NW, Suite 750
Washington, DC 20001
(202) 737-2337
FAX (202) 737-7004
http://www.beerinstitute.org

National Clearinghouse for Alcohol and Drug Information
Substance Abuse & Mental Health Services Administration
11426 Rockville Pike #200
Rockville, MD 20852
(800) 729-6686
http://www.health.org

Distilled Spirits Council of the United States, Inc.
1250 Eye St. NW, #400
Washington, DC 20005
(202) 628-3544
FAX: (202) 628-8888
http://www.discus.org

National Council on Alcoholism and Drug Dependence
20 Exchange Place #2902
New York, NY 10005
(212) 269-7797
FAX: (212) 269-7510
http://www.ncadd.org

National Drug Information and Treatment Hotline
(800) 662-HELP

National Institute on Alcohol Abuse and Alcoholism
6000 Executive Blvd., Willco Bldg.
Bethesda, MD 20892-7003
(301) 443-3860
FAX: (301) 443-7043
http://www.niaaa.nih.gov

Centers for Disease Control and Prevention
National Center for Chronic Disease Prevention & Health Promotion
Office on Smoking and Health
4770 Buford Hwy. NE
Mail Station K-50
Atlanta, GA 30341-3717
(770) 488-5705
(800) 232-1311
FAX: (770) 488-5767
http://www.cdc.gov/tobacco

U.S. Department of Health and Human Services
Substance Abuse and Mental Health Services Administration
5600 Fishers Ln., Room 12-105
Rockville, MD 20857
(301) 443-8956
FAX: (301) 443-9050
http://www.samhsa.gov

Wine Institute
425 Market St., Suite 1000
San Francisco, CA 94105
(415) 512-0151
FAX: (415) 442-0742
http://www.wineinstitute.org

RESOURCES

The various agencies of the U.S. Department of Health and Human Services (HHS) produce important publications on the consumption of alcohol, tobacco, and drugs in the United States and their health effects. Reports of the surgeon general and special reports to Congress are published through this office. The Substance Abuse and Mental Health Services Administration (SAMHSA), an agency of HHS, produces the annual *National Household Survey on Drug Abuse.* HHS also publishes the bimonthly *Public Health Reports,* a helpful resource on health problems caused by alcohol and tobacco.

The National Institute on Alcohol Abuse and Alcoholism (NIAAA), a division of the National Institutes of Health, publishes the periodical *Alcohol Research and Health* (formerly *Alcohol Health and Research World*). The journal contains current, scholarly research on alcohol addiction issues. The NIAAA also publishes the quarterly bulletin *Alcohol Alert,* which disseminates research findings on alcohol abuse and alcoholism.

The National Center for Health Statistics, in its annual *Health, United States,* reports on all aspects of the nation's health, including tobacco- and alcohol-related illnesses and deaths. The Centers for Disease Control and Prevention's (CDC) *Morbidity and Mortality Weekly Report* has published numerous studies on the trends and health risks of smoking and drinking. Additionally, the American Cancer Society and the American Lung Association provide many facts on cancer and heart disease.

The U.S. Department of Agriculture (USDA) is responsible for several helpful reports concerning tobacco. Its publications *Tobacco Situation and Outlook Report and Tobacco: World Markets and Trade* monitor tobacco production, consumption, sales, exports, and imports. The annual *Agricultural Statistics* provides valuable information about farming, and *Food Consumption, Prices, and Expenditures* compiles data on how the nation spends its consumer dollars. The Economic Research Service of the USDA also provides useful data.

The Bureau of Labor Statistics of the U.S. Department of Labor examines how people spend their income, including spending on cigarettes and alcohol.

The National Highway Traffic Safety Administration of the U.S. Department of Transportation produces the annual *Traffic Safety Facts,* which includes data on alcohol-related accidents. The now-defunct Office of Technology Assessment's *Technologies for Understanding and Preventing Substance Abuse and Addiction* (Washington, D.C., 1994) provides a valuable overview of the causes and effects of drug abuse and addiction.

The Bureau of Justice Statistics monitors crime in the United States. Particularly helpful are its releases *Substance Abuse and Treatment, State and Federal Prisoners, 1997* (1999) and *Alcohol and Crime* (1998). The Federal Bureau of Investigation's annual *Crime in the United States* provides arrest statistics for the United States. The Bureau of Alcohol, Tobacco, and Firearms and the Bureau of the Census provide alcohol and tobacco tax information.

The Congressional Research Service prepares reports on various issues for members and committees of Congress, including *Tobacco Master Settlement Agreement (1998): Overview and Issues for the 106th Congress* (1999), *The U.S. Tobacco Industry in Domestic and World Markets* (1998), and *Drunk Driving: Penalties and Incentives Associated with a 0.08 BAC Law* (1999).

Other important annual surveys of alcohol, tobacco, and drug use in the United States are conducted by both public and private organizations. The CDC's *Youth Risk Behavior Surveillance* monitors not only alcohol, tobacco, and drug use, but also other risk behaviors, such as teenage sexual activity and weapons possession. The *Monitoring the Future* survey of substance abuse among students from middle school through college is prepared by the Institute of Social Research of the University of Michigan, and sponsored by the U.S. Department of Health and Human Services. The *PRIDE Questionnaire Report* (National Parents' Resource Institute for Drug Education), based on a survey of youth and parents, is produced by PRIDE Surveys in Bowling Green, Kentucky.

The *Journal of the American Medical Association,* the *American Journal of Psychiatry,* and the *New England Journal of Medicine* are frequent sources of research information on alcohol abuse and addiction. The Agency for Health Care Policy and Research published *Pharmacotherapy for Alcohol Dependence* (1999), outlining the latest pharmacological treatments for dependence and abuse.

The Wine Institute (San Francisco, California), the Distilled Spirits Council of the United States, Inc. (Washington, D.C.), and the Beer Institute (Washington, D.C.) are private trade organizations that track alcoholic beverage sales and consumption, as well as political and regulatory issues. Action on Smoking and Health (ASH) publishes reviews concerned with the problems of smoking and the rights of nonsmokers.

The Gallup Organization (Princeton, New Jersey) and the Robert Wood Johnson Foundation (Princeton, New Jersey), in the *Youth Access to Alcohol Survey,* provide important information about the attitudes and behaviors of the American public.

Information Plus sincerely thanks all of the organizations listed above for the valuable information they provide.

INDEX

recovering alcoholics, 41–42
tobacco, truth in advertising, 93
tobacco use, 51*t*, 52(*t*5.6), 53*t*, 55–57, 56*f*
tobacco use, financial responsibility, 112, 114(*t*10.15)
tobacco use in youth, 66
Heart and heart disease. *See* Cardiovascular system
Heavy drinking, 18*f*, 36*t*
  pregnancy, 24*t*
  youth, 62–65, 65*f*, 73*t*
Henley v. Philip Morris Inc., 99
Hepatitis, 21
Heroin, 3*t*
High-density lipoprotein (HDL), 19
History of alcohol, tobacco, and caffeine use, 7–12, 93
Hospitalization, 20–21

I

Immune system, 23
Imports, tobacco products, 81
Inhalants, 1, 4*t*
Inpatient treatment, 43
Interactions between substances, 24–25, 25*t*, 52, 52(*t*5.6), 104–105, 105*t*
International Classification of Diseases (ICD), 4–5, 35
Internet, 91
Interstate commerce, 91
Intoxication and drunkenness
  history, 8
  physiological aspects, 17–18

K

Ketamine, 2*t*
Kola, 11, 101*t*
Korsakoff's psychosis, 22–23
Kreteks, 50

L

Lawsuits, 94–100
  See also Court cases
Levels of consumption (heavy, light, or moderate)
  alcohol, 16, 26, 26*t*, 27*t*, 28*f*, 36*t*
  tobacco, 46, 49*t*
Liggett Group, Inc., Cipollone v., 94
Liquor
  advertising, 81–85, 86*f*
  distillation, 7
  economics, 75
  per capita consumption, 14*t*
  taxation, 89, 90*t*
  usage polls, 108(*t*10.2)
  youth opinion, 61*t*
  youth use, 71*t*
Liver and liver diseases, 17, 19*f*, 20, 21–22, 21*t*
LSD, 2*t*
Lung cancer. *See* Cancer

M

Malnutrition, 22
MAO (Monoamineoxidase), 47
Marijuana, 2*t*

Master Settlement Agreement (MSA), 97–99, 98*t*, 99*t*
Matching Alcohol Treatments to Client Heterogeneity (Project Match), 43–44
MDMA, 3*t*
Mescaline, 2*t*
MET (Motivational enhancement therapy), 44
Methamphetamine, 3*t*
Methqualone, 2*t*
Methylphenidate, 3*t*
Military, tobacco use, 48–49, 94
Model Statute, 97
Monitoring the Future, 59, 66–68
Monoamineoxidase (MAO), 47
Morphine, 3*t*
Motivational enhancement therapy (MET), 44
Muscle disease, 23

N

Naltrexone, 42
Narcotics, 1, 3*t*
National Center for Health Statistics (NCHS), 15, 47
National Health and Nutrition Examination Survey (NHANES), 15
National Health Interview Survey (NHIS), 15, 47–48
National Household Survey on Drug Abuse, 47
National Institute of Drug Abuse (NIDA), 59
Native Americans, 9
  See also Ethnic and racial factors
Nervous system. *See* Brain and nervous system
Nicotine, 3*t*, 45–47, 57, 93
Norma R. Broin, et al. v, Philip Morris, et al., 99
Novelty drinks, 66

O

Opinions and attitudes
  alcohol and tobacco, 107–114
  liquor advertising, 86*f*, 87*f*
  youth and alcohol and tobacco use, 59–62, 60*t*, 61*t*, 62*f*, 63*t*
Opium, 3*t*
Organized crime, 8
Overdose, caffeine, 104

P

Packaging and labels
  cigarettes, 53
  novelty drinks, 66
Pancreatitus, 22
Parents' Resource Institute for Drug Education (PRIDE), 59
Partner abuse. *See* Domestic violence
PCP, 2*t*
Peer pressure, 4
Per-capita consumption
  alcohol, 13–14, 14*t*, 75
  caffeine, 11–12, 102
  cigarettes, 46*f*, 80
  tobacco, 80, 80(*t*7.7)
  See also Consumption

Philip Morris, et al., Norma R. Broin, et al., v., 99
Philip Morris Inc., Branch-Williams v., 99
Philip Morris, Inc., et al., Estate of Burl Butler, et al. v., 99
Philip Morris Inc. et al., Henley v., 99
Physiological factors
  abuse and addiction, 4
  treating alcoholics, 41–42
Polyabuse, 25
Pregnancy, 17, 23–24, 24*t*, 57, 57*t*, 104
Premature aging, 52
PRIDE (Parents' Resource Institute for Drug Education), 59
PRIDE Questionnaire Report: 1999-2000 National Summary, Grades 6 through 12, 59, 68–69
Production, tobacco, 76–77, 78*t*, 82*t*–84*t*
Prohibition, 1919-1933, 8, 89–91
Project MATCH (Matching Alcohol Treatments to Client Heterogeneity), 43–44
Psilocybin, 2*t*
Psychological factors and effects, 4, 34*t*, 35*t*, 36–37, 43–44
Publicity campaigns, anti-tobacco, 59
Punishment
  drunk driving, 29, 65–66
  drunkenness, 8
  providing alcohol to youth, 112*f*
  youth drinking, 111*f*

Q

Qahwah, 10
Quitting smoking, 56*f*
Quitting tobacco, 111

R

Racial factors. *See* Ethnic and racial factors
Reciprocity (alcohol regulation), 91
Recovery. *See* Treatment and recovery
Regulation
  alcohol, 89–91, 110*t*
  alcohol advertising, 84–85, 86*f*, 87*f*
  tobacco, 89–94, 112–114
  tobacco advertising, 85–86
  See also Bans and restrictions
Reports and surveys
  Diagnostic and Statistical Manual of Mental Disorders (DSM), 4–5
  International Classification of Diseases (ICD), 4–5
  National Health and Nutrition Examination Survey (NHANES), 15
  National Health Interview Survey (NHIS), 15
  National Household Survey on Drug Abuse, 47
  PRIDE Questionnaire Report: 1999-2000 National Summary, Grades 6 through 12, 59
  Smoking and Health: Report of the Advisory Committee to the Surgeon General of the Public Health Service, 53
  Surgeon General reports, 52(*t*5.7)
  Youth Risk Behavior Surveillance—United States 1999, 59

Reproductive system. *See* Sexuality and reproductive system
Respiratory system, 51*t*, 52–54

S

SAMHSA (Substance Abuse and Mental Health Services Administration), 47
Secondhand smoke, 54, 55*t*, 99, 111
Sexuality and reproductive system
    alcohol, effect on, 17, 23
    tobacco, effect on, 51*t*
Smokeless tobacco
    health effects, 51, 66
    history, 9–10
    per capita consumption, 80(*t*7.7)
    youth, disapproval, 63*t*
    youth, perception of harmfulness, 60*t*
    youth use, 50, 64*t*, 68, 69*t*, 70*t*
Smoking. *See* Tobacco and tobacco use
Smoking and Health: Report of the Advisory Committee to the Surgeon General of the Public Health Service, 53
Snuff. *See* Smokeless tobacco
Sobering up, 18
Social factors
    abuse and addiction, 4
    alcohol abuse and addiction, 5*f*, 36, 37*f*
    See also Cultural factors
Soft drinks, 11
Spending. *See* Economics of alcohol and tobacco
State and local government regulation
    alcohol, 91
    tobacco, 94
    See also Regulation
States
    alcohol and tobacco taxes, 90*t*
    blood alcohol concentration laws, 29*t*
    cigarette taxes, 92*f*
    tobacco restrictions in public places, 95*f*, 96*f*
    "Tobacco Wars," 94–99, 99*t*
Statistical information
    alcohol advertising, 86*f*, 87*f*
    alcohol use, 16*t*, 17*t*, 18*f*
    alcohol use, adoption studies, 38*f*
    alcohol use, costs, 42*t*
    alcohol use, crime, 31*t*, 32*t*
    alcohol use, drunk driving accidents, 30*t*
    alcohol use, factors leading to, 37*f*
    alcohol use, health effects, 20(*f*3.5), 21*f*
    alcohol use, levels of consumption, 26*t*, 27*t*, 28*f*, 36*t*
    alcohol use, per capita consumption, 14*t*
    alcohol use, polls, 107*t*, 108*t*, 109*t*
    alcohol use, pregnancy, 24*t*
    alcohol use, public policy, 110*t*
    alcohol use, twin studies, 39*t*
    caffeine, 102*t*
    ethanol consumption, 15*f*
    household expenditures, 76*t*, 77*t*, 80(*t*7.8)

Master Settlement Agreement payments, 99*t*
    taxation of alcohol and tobacco, 90*t*, 91*t*
    tobacco production, 78*t*, 79*t*, 81*t*, 82*t*–84*t*
    tobacco use, cigarettes, 47*f*, 48*f*, 48*t*, 49*f*, 49*t*
    tobacco use, fatalities, 53*t*
    tobacco use, per capita consumption, 46*f*, 80(*t*7.7)
    tobacco use, polls, 113*t*, 114*t*
    tobacco use, pregnancy, 57*t*
    tobacco use, secondhand smoke, 55*t*
    tobacco use, students, 50*t*
    youth, alcohol and tobacco accessibility, 71*t*, 72*t*, 112*f*
    youth, alcohol and tobacco use, 64*t*, 67*t*, 69*t*, 70*t*, 73*t*
    youth, drinking age, 66*f*, 110*f*, 111*f*
    youth, perceptions of alcohol and tobacco use, 60*t*, 61*t*, 62*f*, 63*t*, 65*f*
Steroids, 1, 4*t*
Stimulants, 1, 45, 102
Students. *See* Children and teenagers; College students
Substance Abuse and Mental Health Services Administration (SAMHSA), 47
Surgeons General reports, 52–53, 52(*t*5.7)
Surveys. *See* Reports and surveys
Symptoms of alcoholism and alcohol abuse, 35–36, 35*t*, 36*t*

T

Tar, 51
Taxes and tariffs, 11, 89, 90*t*, 91*t*, 92*f*, 108
Tea, 11, 101, 101*t*, 102*t*
Teenagers. *See* Children and teenagers
Temperance Movement, 1800s, 8, *9*
Tobacco and tobacco use, 45–57
    addiction, 38
    advertising, 85–86
    companies, 94–97, 99–100, 112
    consumer expenditures, 80(*t*7.8)
    history, 7, 9–10, 93
    per capita consumption, 80, 80(*t*7.7)
    production, 76–79, 78*t*, 79*t*, 82*t*–84*t*
    public opinion, 109–114
    regulation, 91–94, 95*f*, 96*f*
    taxation, 89, 90*t*, 91*t*
    usage polls, 113*t*, 114*t*
    youth disapproval, 62*f*, 63*t*
    youth perception of harmfulness, 60*t*, 61*t*
    youth use, 64*t*, 66–68, 67*t*, 69*t*, 70*t*, 73*t*
    youth use, accessibility, 71*t*, 72*t*
"Tobacco Wars," 94–99, 98*t*, 99*t*, 112
Tolerance, nicotine, 45–46
Traffic accidents, 9, 28–29, 30*t*
Treatment and recovery
    alcohol, 41–44
    drugs, 6*f*
    tobacco, 55–57
Twelve-step programs, 43–44
Twenty-first Amendment, 8, 89–91

Twin studies, 39–40, 39*t*
Types A and B alcoholism, 40, 40*t*, 41*t*

U

United Kingdom, 11
United States, early history. *See* Colonial America
Uric acid, 102
USFDA. *See* Food and Drug Administration (FDA)

V

Violence and crime, 29–32, 31*t*, 32*t*

W

Warning labels. *See* Packaging and labels
Wernicke's disease, 23
WHO. *See* World Health Organization (WHO)
Wholesalers, alcohol, 91
Wine
    advertising, 87(*t*7.2)
    economics, 75
    history, 7
    per capita consumption, 14
    taxation, 89, 90*t*
    usage polls, 108(*t*10.2)
    youth opinion, 61*t*
Withdrawal symptoms
    caffeine, 103
    tobacco, 46, 57
Women's issues
    alcohol abuse and addiction, 38, 40
    alcohol use, 14, 16*t*, 17*t*, 19, 22
    alcohol use during pregnancy, 24*t*
    caffeine, 104
    cirrhosis of the liver, 22
    fetal alcohol syndrome, 23–24
    former smokers, 56–57
    secondhand smoke, 54
    tobacco advertising, 85
    tobacco use during pregnancy, 57*t*
    youth smokers, 68
World Health Organization (WHO), 4, 35
Worldwide alcohol, tobacco, and caffeine issues
    antismoking campaigns, 55
    caffeine farming, 101*t*
    history of alcohol use, 8
    tea consumption, 11
    tobacco imports and exports, 81, 81*t*
    tobacco production, 11, 82*t*–84*t*
    tobacco use, 9–10
    wine consumption, 75

Y

Youth Risk Behavior Surveillance—United States 1999, 59, 68

Z

Zyban, 57